André Thess

The Entropy Principle

Thermodynamics for the Unsatisfied

André Thess

The Entropy Principle

Thermodynamics for the Unsatisfied

With 55 Figures and 4 Tables

Springer

Author

Prof. Dr. André Thess

Institute of Thermodynamics and Fluid Mechanics
Ilmenau University of Technology
P.O. Box 100565
98684 Ilmenau
Germany

ISBN 978-3-642-13348-0

Library of Congress Control Number: 2010936708

Typesetting: Stasch, Bayreuth (stasch@stasch.com)
Language editing: Stephanie Polster, USA
Production: Heather King

Printed on acid-free paper

9 8 7 6 5 4 3 2 1

springer.com

Temperature … is a corollary of entropy;
it is epilogue rather than prologue.

Elliott Lieb, Jakob Yngvason

Preface

Is it possible to define entropy in classical thermodynamics in a way that is mathematically accurate and at the same time easy to understand? I have often asked this question to myself first when I was a student and later when I became a professor of mechanical engineering and had to teach thermodynamics. Unfortunately, I never got a satisfactory answer. In textbooks for physicists I often found the claim that entropy can only be "really" understood when one has recourse to statistical physics. But it appeared strange to me that a physical law as perfect as the second law of thermodynamics, which is closely related to entropy, should depend on tiny details of the molecular structure of the matter that surrounds us. By contrast, in textbooks for engineers entropy was most often defined on the basis of temperature and heat. However, I never felt comfortable with the idea that such a fundamental quantity as entropy should be determined on the basis of two concepts which cannot be accurately defined without entropy. Given this state of affairs, I came close to resignation and was on the verge of believing that an accurate and logically consistent definition of entropy in the framework of a macroscopic theory was altogether impossible.

In the spring of the year 2000, I came across an article entitled "A Fresh Look at Entropy and the Second Law of Thermodynamics" written by the physicists Elliott Lieb and Jakob Yngvason which appeared in the journal *Physics Today*. Their idea that the concept of adiabatic accessibility rather than temperature or heat is the logical basis of thermodynamics appealed to me immediately. For the first time in my academic life I began to feel that I really understood the entropy of classical thermodynamics. However, it took considerable effort to study and understand the article "The Physics and Mathematics of the Second Law of Thermodynamics" (*Physics Reports*, vol. 310, 1999, pp. 1–96) by the same authors in which the full "Lieb-Yngvason theory" is presented. Once I had finished the work, however, I was convinced that the Lieb-Yngvason theory represents the ultimate formulation of classical thermodynamics. Although the theory is mathematically complex, it is based on an idea so simple that each student of science or engineering should be able to understand it.

I then decided to involve my students in order to test whether the Lieb-Yngvason theory is as convincing as I believed. I have been teaching a one-year thermodynamics course for the undergraduate mechanical engineering students of Ilmenau University of Technology since 1998. I use Moran and Shapiro's textbook "Fundamentals of Engineering Thermodynamics" (Wiley and Sons), and I introduce entropy as is currently most often done in engineering courses, namely via the Carnot process cycle and the Clausius inequality. One week after having introduced entropy in the regular lecture,

I invited the students to a voluntary 90-minute supplementary lecture and presented them with the basic ideas of the Lieb-Yngvason theory. I was prepared for the worst-case scenario and predicted that the roughly sixty students who showed up would shower me with a torrent of rotten eggs and tomatoes after being confronted with my presentation of this complex mathematical apparatus. Surprisingly, the opposite happened. In an anonymous questionnaire that I distributed after the lecture, more than two-thirds of the students stated that they found the definition of entropy according to the Lieb-Yngvason theory more plausible than the traditional one. Some even asked me if I could incorporate the Lieb-Yngvason theory into my regular thermodynamics course.

Such an encouraging response prompted me to extend and write up my lecture notes in order to make Lieb and Yngvason's work accessible to a broader readership. This book is intended as supplementary material to existing textbooks of thermodynamics and is primarily written for those students who find the traditional definitions of entropy unsatisfactory. Accordingly, the contents of the book do not constitute a self-contained introduction to thermodynamics. It is rather suited as a short course on the accurate formulation of classical thermodynamics whose content can be adjusted to the time and interests of the students. In my experience, the minimal version of a supplementary short course consists in a single lecture lasting 90 to 120 minutes which is presented in addition to the regular thermodynamics course. In this time, the lecturer can cover the basic ideas of Chap. 1, 2 and 3. However, in later years it turned out to be more appropriate to use two or three hours. The first half of the time can be devoted to the fundamentals explained in Chap. 1–3. During the second part, Chap. 4 as well as one or two selected examples from Chap. 5 can be covered.

I hope that this book conveys some of the fascination that I experienced when studying the Lieb-Yngvason theory. I would like to emphasize that I have made no contribution to the formulation of the theory presented here. My only task was to translate the mathematically complex theory into the language of undergraduate science and engineering students. Accordingly, I take the responsibility for all errors or misrepresentations of the original theory. The Lieb-Yngvason theory itself, I believe, is perfect.

It is my pleasure to express my gratitude to Elliott Lieb and Jakob Yngvason for answering many questions and encouraging me to write this book as well as prepare an English version. Moreover, I would like to thank Friedrich Busse, Gerhard Diener, Walter John and Holger Martin for valuable suggestions and useful discussions. I also thank Rainer Feistel, Achim Dittmann, Andreas Pfennig and Roland Span for their help with thermodynamic data. Finally, my thanks go to Cornelia Gießler, Martina Klein and Renate Heß for their help in preparing the figures and to Armin Stasch, scientific book designer, for his invaluable help with typesetting.

The authors and the date of origin of the Holy Bible are unfortunately unknown. For the bible of classical thermodynamics, however, this information is readily available: It was written by Elliott Lieb and Jakob Yngvason, it carries the title "The Physics and Mathematics of the Second Law of Thermodynamics" and it was first published in 1999 in the journal *Physics Reports*. I hope that this book will encourage its readers to study the bible of thermodynamics!

André Thess
Ilmenau, 18 January 2010

Contents

Chapter 1 Introduction

Chapter 1

Introduction

1.1
Is Entropy Really Necessary?

Why does a glacier flow? Why does a diver die when he rises too quickly to the surface of the water? Why does salt pull the water out of a cucumber? Why does a warm bottle of soda produce more foam upon opening than a cold one? Why is it impossible to use the heat of the Gulf Stream for the production of electricity? All these questions have one property in common: they cannot be answered without considering entropy.

Temperature, pressure and volume are quantities of which we have a sound intuitive understanding. Internal energy U, whose existence is postulated by the first law of thermodynamics, is slightly more difficult to comprehend. Yet everyone becomes familiar with this quantity after solving a couple of textbook problems. But how about entropy? Many people confess that they have never understood the real meaning of this intriguing quantity. Do we need it at all? Can't we do thermodynamics without struggeling with entropy? Are the foregoing quantities not sufficient to solve all practical thermodynamic problems?

A simple example shows us that this is not the case. When a stone falls into a well, the energy of the water increases by the same amount as the decrease in potential energy of the stone. Can this process run spontaneously in the reverse direction? It would certainly not violate the first law of thermodynamics if water would spontaneously cool down a little bit and conspire to throw a stone into the sky. Nevertheless, our experience shows that such a process doesn't ever occur. Our physical intuition suggests that "something" has been lost in the system consisting of the water and the stone once the stone has fallen down. In what follows we will see that this "something" can be accurately described in mathematical terms and leads us to the concept of entropy. Thanks to a recent work by the physicists Elliott Lieb and Jakob Yngvason (Lieb and Yngvason 1999), we will be able to give a definition of entropy which is characterized by unprecedented mathematical and logical rigor. The purpose of this book is to make this "Lieb-Yngvason theory" accessible to undergraduate students as well as to other scientists and engineers interested in the fundamentals of thermodynamics.

The logical chain of arguments that lead us to entropy starts with the order relation

$$\prec \tag{1.1}$$

called adiabatic accessibility. The overarching goal of the present book is to understand how entropy can be derived from this fundamental concept. To give the reader a first glance at the result, the entropy S of the state X of a thermodynamic system is given by the relation

$$S(X) = \max\{\lambda : ((1-\lambda)X_0, \lambda X_1) \prec X\} \tag{1.2}$$

This definition is awkward, as it contains unusual symbols and does not appeal to our intuition. Few readers, if any, will find that something has been gained by replacing a seemingly clear definition like $dS = \delta Q / T$ (which can be found in many textbooks of thermodynamics) by Eq. 1.2. However, it turns out that Eq. 1.2 is only slightly more complicated than the formula

$$W(X) = \min\{\lambda : \lambda Y_0 \prec X\} \tag{1.3}$$

defining the value W of an object X in the fairy tale "Hans in Luck". Let us turn to this popular piece of German literature, written by the brothers Grimm, in order to acquire the mathematical tools necessary to understand entropy.

1.2 A Didactic Model for the Logical Structure of the Entropy Principle

> Hans had served his master for seven years. When he left, his master gave him a piece of gold in reward for his work. Soon after, when Hans set out on his way home, the gold started to hurt his shoulder. He was happy to find a horseman who gave him his horse in exchange for the gold. But the horse threw him off, so he gave it away for a cow, the cow for a pig, the pig for a goose and the goose for a grindstone. When he stopped at a well in a field to refresh himself, he slipped and fell, pushing against the grindstone so that it fell into the water. He thanked God for having delivered him in such a good way from these heavy burdens which had been the only things that troubled him. "There is no man under the sun so fortunate as I" he cried out. With a light heart and freedom from every burden he now ran uninhibited until he arrived home where his mother was waiting.

The full fairy tale is given in the Appendix A.

Parents read this fairy tale to their children in order to teach them that each object has a value and that this value can be irrevocably destroyed by careless action. The parents could equally well show Eq. 1.3 to their children but this is unlikely to be a good idea. We, however, would like to use the (incomplete) analogy between the value W in Grimm's fairy tale and the entropy S in physics to convince ourselves that entropy is not as difficult to understand as Eq. 1.2 might suggest. The logic behind the derivation of entropy in Chap. 2 and 3 will turn out to be similar to the definition of the value in the present chapter.

Let us denote Hans' property by the symbol X, which stands for G (gold), H (horse), C (cow), P (pig), B (goose – with B from bird), S (grindstone – with S from stone), O (nothing). In addition we shall assume that Hans could have exchanged the piece of gold for a house R (with R from residence) and could have reversed this deal at any time. We now define financial accessibility as a basis for understanding adiabatic accessibility in Chap. 2.

Financial accessibility. An object Y is financially accessible from an object X, denoted $X \prec Y$ (pronounced "X precedes Y"), if there is a market in which it is possible to exchange X for Y.

In our example we obviously have $G \prec H$, $H \prec C$, $C \prec P$, $P \prec B$, $B \prec S$, $S \prec O$. Since we have made the additional assumption that Hans could have exchanged his wages for a house R and the house back for the gold, both $G \prec R$ and $G \succ R$ (pronounced "G succedes R") must hold. We shall denote two objects which are mutually financially accessible as *financially equilvalent*, and write $G \stackrel{E}{\sim} R$. If Y is accessible from X but the converse is not true, we shall write $X \prec\prec Y$ (pronounced "X strongly precedes Y"). In our fairy tale, there is $G \prec\prec H$, $H \prec\prec C$, $C \prec\prec P$ and so on. We further assume that a spontaneous gain which would correspond to exchanging a grindstone for a piece of gold is impossible. An important property of the relation $\prec$ should be particularly emphasized. The decision whether the statement $X \prec Y$ is true is a binary decision. This means that the answer to the question "Is $X \prec Y$ true?" must be either **YES** or **NO**. Consequently, the information about the financial accessibility of a given set of objects can be represented by a table with binary entries.

a

	G	H	C	P	B	S	0	R
G	Y	Y	Y	Y	Y	Y	Y	Y
H	N	Y	Y	Y	Y	Y	Y	N
C	N	N	Y	Y	Y	Y	Y	N
P	N	N	N	Y	Y	Y	Y	N
B	N	N	N	N	Y	Y	Y	N
S	N	N	N	N	N	Y	Y	N
0	N	N	N	N	N	N	Y	N
R	Y	Y	Y	Y	Y	Y	Y	Y

b

$W(G) = 1$
$W(H) = 0.8$
$W(C) = 0.7$
$W(P) = 0.5$
$W(B) = 0.3$
$W(S) = 0.2$
$W(0) = 0$
$W(R) = 1$

Fig. 1.1. *Hans in Luck and the value principle.* Illustration of the financial accessibility (a) and the hypothetic value (b) of objects in the brothers Grimm's fairy tale "Hans in Luck". The abbreviations denote gold (G), horse (H), cow (C), pig (P), goose (B) (like **b**ird), grindstone (S), and nothing (0). For illustrative purposes, an additional object has been introduced, a house (R) (like **r**esidence), which is not present in the fairy tale. In table (a) **Y** and **N** denote respectively **YES** and **NO** as answers to the question whether an object in the first line of the table is financially accessible from an object in the first column. For instance, the response to the question "Is $G \prec S$ true?" is **YES** (line 2, column 7), whereas the answer to "Is $S \prec G$ true?" is **NO** (line 7, column 2). The units of the values in (b) are arbitrary. The numbers do not have a particular meaning except that they roughly reflect the order of the values of the considered objects

Figure 1.1 shows the result that we would have obtained if we had checked the financial accessibility of all objects occurring in "Hans in Luck". Although our example contains as little as eight different objects, the table is already quite large. Is it possible to represent the contents of this table in a more concise manner? This is indeed the case if we introduce the notion of *value*. Figure 1b shows one of many possibilities of assigning to each object X a value $W(X)$. The information content of both tables is the same. However, Fig. 1b is much more compact than Fig. 1a. If we wish to know whether Y is financially accessible from X we only need to compare the values $W(X)$ and $W(Y)$. If $W(X) > W(Y)$, then $X \prec\!\prec Y$, whereas if $W(X) < W(Y)$, then $X \succ\!\succ Y$. Finally if $W(X) = W(Y)$, then obviously $X \prec Y$ and $X \succ Y$ and thus $X \stackrel{\pounds}{\sim} Y$. The conclusion that the relation $\prec$ of financial accessibility between different objects can be completely described – or encoded – by a function $W(X)$ can be mathematically expressed in the following way.

Value principle. There is a real valued function on all objects X called value and denoted by W such that:

- Monotonicity: If $X \prec\!\prec Y$ then $W(X) > W(Y)$, if $X \stackrel{\pounds}{\sim} Y$ then $W(X) = W(Y)$
- Additivity: $W((X, Y)) = W(X) + W(Y)$

In contrast to the entropy principle discussed in later chapters, the value principle should not be considered as a rigorous mathematical result but rather as an illustrative informal example. It can be readily verified that the values given in Fig. 1b satisfy these properties. The monotonicity ensures that two objects are financially equivalent if their values are equal and that a series of careless exchanges with $X \prec\!\prec Y$ and $Y \prec\!\prec Z$ is equivalent to a continuous degradation of values with $W(X) > W(Y) > W(Z)$. Additivity implies that the value of a compound system (X, Y), i.e., a system which contains two objects X and Y, is equal to the sum of the individual values. This property seems trivial. But this is not the case. Indeed, additivity of W has far-reaching consequences. For instance, additivity implies that states are financially accessible for a compound system which is not financially accessible for a single object. To give an example, two geese ($Y = 2B$) are financially inaccessible from one goose ($X = 1B$) because $W(1B) < W(2B)$. If, however, we have additional resources at our disposal, – for instance two pieces of gold, – then the compound system $(1B, 2G)$ can financially access the compound system $(2B, 1G)$ because the value $W(1B, 2G) = 2.3$ of the former exceeds the value $W(2B, 1G) = 1.6$ of the latter. The additivity, like an invisible band, connects objects with entirely different properties. A haircut and a hamburger have very little in common, except that both have a certain value.

After having acquainted ourselves with the properties of the function $W(X)$, we turn to the question how to determine it. In the first step, we choose an arbitrary good Y_0 as a reference object. For instance, a piece of gold with a mass of one kilogram could serve as such an object. In the second step, we introduce a scaling factor λ and denote λY_0 as a *scaled copy* of Y_0. In our example $\lambda = 0.3$ corresponds to 300 grams of gold. In a third and final step we define the value of an object as the minimum of λ for which the quantity λY_0 of gold is sufficient in order to financially access the object. The mathematical expression of this definition is

$$W(X) = \min\{\lambda : \lambda Y_0 \prec X\} \tag{1.4}$$

This definition should be pronounced as follows: "The value of X is equal to the minimum of λ which has the property that X is financially accessible from λY_0." The so-defined value is dimensionless. If $\lambda = 2$ holds for an object, then its value is equal to the value of two kilograms of gold. In practice, however, we use dimensional values. We transform $W(X)$ into a dimensional quantity by allocating a unit, say 10 000 €, to the reference object. We then obtain the dimensional value as

$$W_*(X) = W(X) \times 10000\,€ \tag{1.5}$$

Thus, the determination of the value according to the Definition 1.4 is reduced to a series of experiments, each of which yields a yes-or-no answer. In our example, we could determine the value in two ways by approaching the minimum of λ either from below or from above. A generous person would first exchange one kilogram of gold ($\lambda = 1$) for a goose, notwithstanding the horrendous financial loss associated with this transaction. She or he would then repeat the experiment by decreasing the amount of gold in steps of, say, one gram. The 702th experiment (question: "$0.299\,Y_0 \prec X$?", answer: "NO!") would be unsuccessful and thus the value would have been determined as $W = 0.3$. Conversely, a mean person would start with one gram of gold ($\lambda = 0.001$). After 299 unsuccessful attempts to exchange amounts of gold increasing in steps of one gram, the affirmative answer to the question "$0.3\,Y_0 \prec X$?" would lead him or her to the conclusion that $W = 0.3$. We thus realize that the value of an object can be determined in a strictly deterministic way. Equipped with such a profound understanding of German literature we can summarize our findings in the following sentence (which should not be taken too seriously).

> **Principle of value decrease.** In the hands of a fool property will never increase its value, i.e., $\Delta W \leq 0$.

With this statement, we finish our discussion of the didactical model which should be considered as an informal introduction rather than a rigorous mathematical theory. In order to prepare ourselves for the return to thermodynamics, we summarize our conclusions regarding financial accessibility and value in the following compact form.

1. Objects can be sorted using the order relation $\prec$, which we denote as financial accessibility;
2. The object Y is said to be financially accessible from the object X, written $X \prec Y$, if there is a market in which X can be exchanged for Y;
3. Our everyday experience about the financial accessibility of objects can be summarized in the form of the following value principle.
4. **Value principle.** There is a real valued function on all objects X called value and denoted by W such that W is monotonic [if $X \prec\prec Y$ then $W(X) > W(Y)$, if $X \overset{E}{\sim} Y$ then $W(X) = W(Y)$] and additive [$W((X, Y)) = W(X) + W(Y)$].

Before turning to the following chapter it is instructive to take a quick look at the summary of the whole book which is given in Chap. 6. This will lead us to the conclusion that with our introductory example we have already made a great step towards understanding the entropy principle. Ahead of us is a lot of technical work, but the logic behind this work will be very similar to what we have just learned.

Chapter 2 Adiabatic Accessibility

Adiabatic Accessibility

2.1 Thermodynamic Systems

Thermodynamics is the science of the transition of thermodynamic systems between different equilibrium states and the associated changes of their energy and chemical composition. A thermodynamic system is a well-defined quantity of matter which is scalable and which can exchange energy with its environment. Examples of thermodynamic systems are 1 kg of ice, 500 g of alcohol, 450 g of red wine or 1 mole of sulfuric acid. *Scalability* refers to the possibility of splitting the system into several parts without changing the properties or recombining several systems into a larger one. For instance, we can divide 450 g of red wine among three glasses with 150 g each without changing the properties of the wine. By contrast, a single molecule does not constitute a thermodynamic system because it lacks the property of scalability. Indeed, if we let two molecules react and form one large molecule, the properties of the latter will be different from its constituents. Our Universe is not a thermodynamic system either, since its division or duplication is unlikely to make physical sense. Complex systems like humans or animals do not represent thermodynamic systems either, because it is not possible to split them into living small sub-individuals. Neither is it possible to combine one million mice to create a single giant living mouse. In summary, a thermodynamic system should be neither too small nor too large or complex.

2.2 Equilibrium States

Classical thermodynamics deals with systems that are in the state of thermodynamic equilibrium. These states will be referred to as equilibrium states or simply *states*. We shall define them as follows.

> **Equilibrium states** (physical definition). A thermodynamic system is said to be in an equilibrium state if its properties are stable in time and can be described by a small number of macroscopic quantities.

The exploding boiler of a steam locomotive, for instance, is not in a thermodynamic equilibrium state because the process proceeds very fast and the number of data necessary to specify the details of the instantaneous motion of the two-phase mixture and the debris is huge. By contrast, 500 grams of Vodka in a refrigerator is very close to an equilibrium state, which is characterized by the temperature, pressure and alcohol content and which is relatively stable in time.

The given explanation of equilibrium states was necessary in order to gain a clear intuitive understanding of the systems to be analyzed. For the mathematical formulation of thermodynamics, however, none of these issues is relevant. From a mathematical viewpoint, the equilibrium states which we shall denote as X, Y, Z or A, B, C are merely elements of an abstract *state space*. For instance, X (1 kg ice at the melting point at atmospheric pressure) and Y (1 kg water at the boiling point at 10 bar) are two elements of the state space of one kilogram of water. Scalability implies that the state λX (where $\lambda = 2$ is a scaling factor) represents a state belonging to the state space of two kilograms of water. Moreover, we assume that we can always split a system that is in the state X into two parts according to $(\lambda X, (1-\lambda)X)$ with $0 \leq \lambda \leq 1$, or we can recombine two parts into a single system.

Whether a given system is in an equilibrium state is a question outside the realm of thermodynamics. Similarly, classical mechanics cannot answer the question whether Earth is a point mass whose state is uniquely determined by its position and velocity in the three-dimensional Euclidian space. In thermodynamics, one always has to check whether the assumption of an equilibrium state is justified by comparing the result of a computation with an experiment. In a similar way, the Earth can be treated as a point mass if its trajectory computed under this assumption agrees with the true one. Before we turn to the discussion of adiabatic accessibility, we summarize the mathematical content of this paragraph as follows.

Equilibrium state (mathematical formulation). The equilibrium states X, Y, Z of a thermodynamic system are elements of an abstract set termed state space.

2.3 The Order Relation $\prec$

The order relation $\prec$, which is called *adiabatic accessibility*, plays a central role in paving the way to the definition of entropy. From now on, no further reference will be made to the notion of financial accessibility that was used in our introductory example. The symbol $\prec$ will be exclusively used for adiabatic accessibility. Following Lieb and Yngvason (1999), we define:

Adiabatic accessibility (physical formulation). A state Y is adiabatically accessible from a state X, in symbols $X \prec Y$ (to be pronounced "X precedes Y" or "Y succeeds X"), if it is possible to change the state from X to Y by means of an interaction with some device (which may consist of mechanical and electrical parts as well as auxiliary thermodynamic systems) and a weight, in such a way that the device returns to its initial state at the end of the process (or can be returned), whereas the weight may have changed its position in a gravitational field.

Let us illustrate the meaning of adiabatic accessibility using some specific examples. Figure 2.1a shows a thermodynamic system which consists of a certain amount of water.

The system undergoes a transition from the state X describing ice at its melting point under atmospheric pressure to the state Y describing water at its boiling point under atmospheric pressure. The transition is accomplished by two devices, namely a microwave oven and a generator. At the end of the process, both devices return to their initial states, whereas the weight which drives the generator has lowered its position. It is thus obvious that the relation $X \prec Y$ holds. A transition from a state X to another state Y whose sole effect in the environment is the change of the position of a weight shall be denoted as *adiabatic process*. An adiabatic process is uniquely characterized by its initial state X, final state Y and the work W that has been performed on the system. In Fig. 2.1a we could have accomplished the melting of the ice in a much simpler way, namely by mechanical stirring. We have instead chosen to include a generator and a microwave oven in our example in order to make clear that the device used to execute the adiabatic process can be quite complex.

Can we reverse the process shown in Fig. 2.1a? Is Y adiabatically accessible from X? A spontaneous transition from Y to X in the form of an adiabatic process is obviously impossible. It would certainly not violate the principle of energy conservation if hot water would spontaneously freeze and the released energy would either be transformed into electricity or used to lift a weight. However, such a process has never been observed. If the process were possible, all of the energy problems in the world would immediately evaporate. Indeed, one could produce electricity *ad libitum* by cooling the oceans. Their enormous mass would be sufficient to satisfy all current and future needs of mankind. Since we have no indication that such a miracle is possible, we arrive at the conclusion that it is impossible to transform the considered system from Y to X by an adiabatic process. Mathematically, this conclusion can be written as $Y \nprec X$. In what follows we shall denote states which obey $X \prec Y$ and $Y \nprec X$ as $X \prec\prec Y$ (pronounced "X strongly precedes Y").

At this point, the reader could justifiably ask whether it would not be possible to find other ways to adiabatically transform the system from state Y back into state X. For instance, one could imagine using liquid nitrogen to freeze the water. In this case, our "device" would consist of liquid nitrogen. However, since nitrogen transforms from a liquid to a gaseous state during our process, our device does not return to its initial state and the condition for an adiabatic process is not fulfilled. Couldn't we build a second device which brings the evaporated nitrogen back into its initial state? As a matter of fact, the experience of all scientists and engineers tells us that this is not possible. This example shows us that the order relation $\prec$ reflects our everyday experience that it is impossible to reverse certain processes in nature and engineering.

Figure 2.1b shows another example of an adiabatic process. This time our example involves a compound system consisting of two cups of water in the initial state X. If we connect the hot and the cold cup with a copper wire, we observe that their temperatures will converge. After a sufficiently long time, they will become equal and no further change will occur. We can then remove the wire representing our "device" and observe that it has returned to its initial state. This demonstrates that the considered transformation to the final state denoted by Y is an adiabatic process and that $X \prec\prec Y$ holds. It should be emphasized that the notion of adiabatic accessibility is only defined for equilibrium states. During the period when the two subsystems exchange energy, our system is not in an equilibrium state. It is then invisible to thermodynamics very much like a mole which only rarely appears on the surface of the Earth. Since the

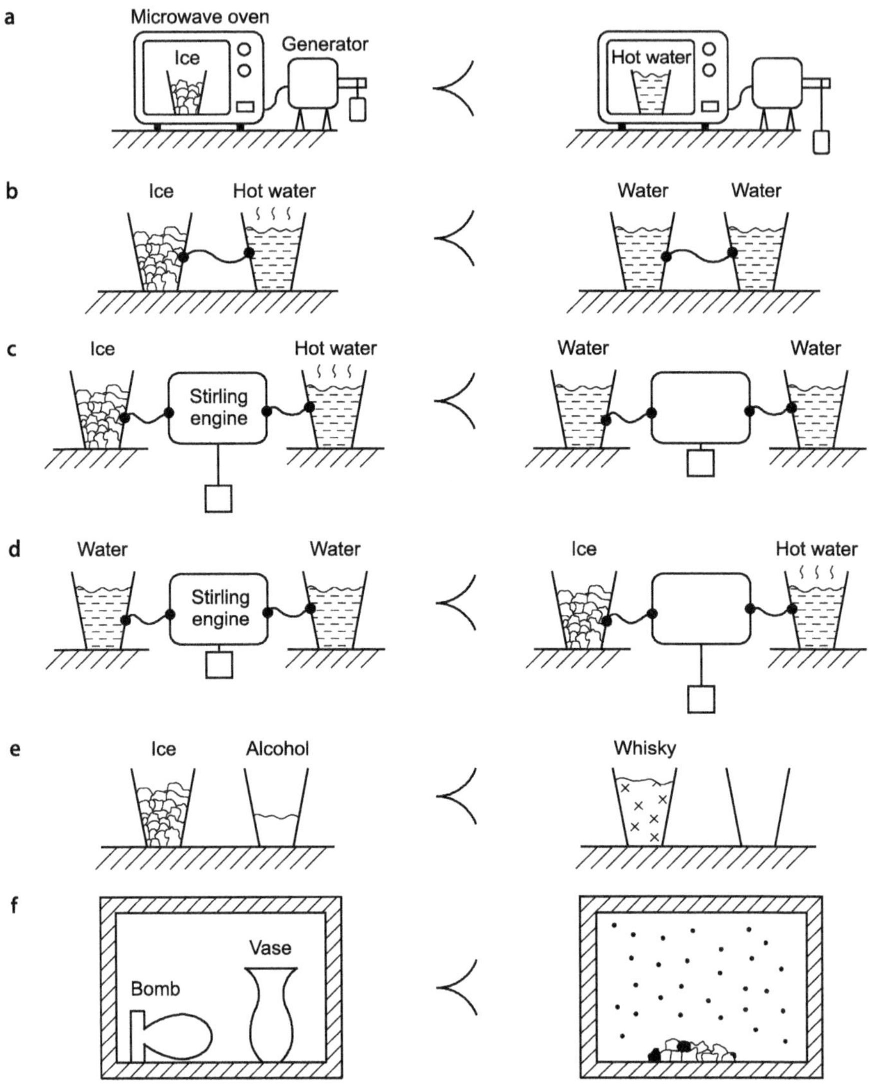

Fig. 2.1. *Examples of adiabatic accessibility.* Equilibrium states of thermodynamic systems before and after an adiabatic process. In all examples, the initial and final states are denoted as X and Y, respectively. (a) Melting of ice in a microwave oven characterized by $X \prec Y$ and $X \not\succ Y$ (which can also be written as $Y \not\prec X$) and thus $X \prec\prec Y$. (b) Thermal equilibration by heat conduction characterized by $X \prec\prec Y$. (c) Thermal equilibration using a Stirling engine which performs work by lifting a weight. In general, $X \prec\prec Y$ but in the particular case of an ideal engine, discussed in Sect. 4.5 A, $X \prec Y$ as well as $Y \prec X$ and thus $X \overset{A}{\sim} Y$. (d) Generation of a temperature difference using a refrigeration system which receives mechanical energy from a weight lowering its position. In general, $X \prec\prec Y$ but in the particular case of an ideal engine, discussed in Sect. 4.5 B, $X \overset{A}{\sim} Y$ as in example (c). (e) Mixing process with $X \prec\prec Y$. (f) Example of a very rapid adiabatic process with $X \prec\prec Y$. During the transition from X to Y, the systems (a) through (f) are *not* in general in an equilibrium state. The reverse processes to (a), (b), (e), and (f) are never adiabatic. The reverse processes (c) and (d) are only adiabatic in the case of ideal engines

spontaneous reversal of the process shown in Fig. 2.1b is apparently impossible, we have $Y \not\prec X$, which together with $X \prec Y$ leads to $X \prec\prec Y$.

Before we proceed with the explanation of examples, let us briefly discuss the relation between the notion of adiabatic accessibility as defined by Lieb and Yngvason 1999 and the traditional concept of adiabatic processes. In the traditional understanding, a process is called adiabatic if the system is thermally insulated from its environment and cannot exchange heat with its surroundings. However, as we shall see in Chap. 3 and 4, *temperature* and *heat* cannot be accurately defined without entropy. It is therefore preferable to avoid the use of such concepts until entropy has been introduced. By contrast, adiabatic accessibility does not rely on concepts like temperature, heat or thermal insulation. Not even words like *hot* or *cold* are necessary to define adiabatic accessibility. Indeed, we could have characterized the energy exchange in Fig. 2.1b by purely mechanical or electrical means, for instance by using thermocouples.

Figure 2.1c shows that the flow of energy between two thermodynamic systems is not necessarily as useless as in example b. Indeed, it is possible to produce work as a result of energy exchange. If we bring the ice and the hot water in contact with a heat engine, for instance a Stirling engine, we can perform work. We will discuss such processes in detail in Sect. 4.5 and 5.4. At this point, it shall be sufficient to say that in the general case $X \prec\prec Y$ holds for processes such as shown in Fig. 2.1c. However, there are ideal heat engines which can accomplish the transformation in such a way that both $X \prec Y$ and $X \succ Y$ are true. In such a case, we call the states X and Y adiabatically equivalent and write $X \stackrel{A}{\sim} Y$. We parenthetically note that in the case of an ideal cycle, the temperature of the water in the state Y_c, shown in Fig. 2.1c, is lower than in Y_b, shown in Fig. 2.1b and that $Y_c \prec\prec Y_b$.

Figure 2.1d illustrates that an adiabatic process does not necessarily lead to a reduction of temperature differences between two thermodynamic systems. For instance, a refrigeration system can transform two glasses of water from an initial state with equal temperatures to a final state consisting of ice and hot water. (We will discuss the principles of operation of refrigeration systems in Sects. 4.5 and 5.5.) Such transformation requires work which is symbolically expressed by a weight lowering its position in the gravity field of the Earth. Similar to the example from Fig. 2.1c, $X \prec\prec Y$ holds for the considered adiabatic process in general. However, we can imagine an ideal refrigeration system which accomplishes the process in such a way that both $X \prec Y$ and $X \succ Y$ hold and thus $X \stackrel{A}{\sim} Y$.

At first glance, it may seem that it is possible to adiabatically reverse the process shown in Fig. 2.1b by using a refrigeration system as in example d. However, this is not the case. With the current level of our understanding, we are not able to prove this. After having introduced entropy, we will be able to demonstrate that this process cannot be adiabatically reversed.

The process shown in Fig. 2.1e obviously has the property $X \prec\prec Y$ but is more complex than the previous examples because it involves mixing. The initial state is a member of the state space of say 100 g water and 50 g alcohol, whereas the final state is a member of the state space of 150 g Vodka. This example shows us that the structure of the state space can change in the course of an adiabatic process. This change of state space is one of the reasons for the mathematical complexity of many real-life problems of thermodynamics.

As a final example, Fig. 2.1f demonstrates that the concept of adiabatic accessibility should not be confused with slow or quasi-static processes. "The explosion of a bomb in a closed container is an adiabatic process" (Lieb and Yngvason 1999). The list of examples in Fig. 2.1 is by far not exhaustive. We shall soon consider more adiabatic processes, such as for instance the compression of a gas.

Let us briefly mention a subtlety in the foregoing physical definition of adiabatic accessibility. There are adiabatic processes where the device does not return to its initial state. For instance, when a gas is compressed in a cylinder using a piston, the device consisting of the cylinder, the piston and perhaps auxiliary equipment, does not come back to its initial state because the piston has changed its position. In such a case, it is important whether it is possible *in principle* to return the device to its initial condition in a purely mechanical manner. We have added the words "or can be returned" in the definition of adiabatic accessibility in order to take into account this possibility.

The physical explanation of the concept of adiabatic accessibility and the discussion of the examples in Fig. 2.1 are necessary to develop a clear intuitive understanding of the phenomena that we wish to analyze. For the mathematical formulation of thermodynamics, however, these pictures are not necessary. From a mathematical point of view, the following statement is important:

> **Adiabatic accessibility** (mathematical formulation). The adiabatic accessibility $\prec$ is an order relation between the equilibrium states X, Y, Z of a thermodynamic system.

Equilibrium state and *adiabatic accessibility* are so-called primitive concepts of thermodynamics. Their mathematical content cannot be reduced to more fundamental concepts. In a similar way, *point* and *event* are primitive concepts of geometry and probability theory, respectively. Thus, equilibrium states and adiabatic accessibility are the fundamentals of thermodynamics from which the existence of entropy and the derived concepts of temperature and heat follow. Thermodynamics is often presented in the opposite order. But we would like to emphasize that this is logically not correct.

2.4 A First Glance at Entropy

In Chap. 3 we will explain in detail how entropy is constructed. This explanation will require some mathematical tools which we have not defined yet. But even without these tools we are already able to understand why the introduction of entropy greatly simplifies the description of adiabatic accessibility between equilibrium states.

Let us consider Fig. 2.2 which shows a gas in a thermally insulated container. We assume that the gas represents our thermodynamic system whereas the container, the piston and the other elements, except for the weights, represent the device in the sense of our definition of adiabatic accessibility. First we analyze the case which is represented in Figs. 2.2a and b, where the container is closed by a piston which can be moved up and down. We assume that our system is transformed from the initial state X shown in Fig. 2.2a to the final state Y sketched in Fig. 2.2b. This process is accomplished using a device which places a weight onto the piston. It is not important for our considerations how this device is actually constructed. We shall only assume that it derives its

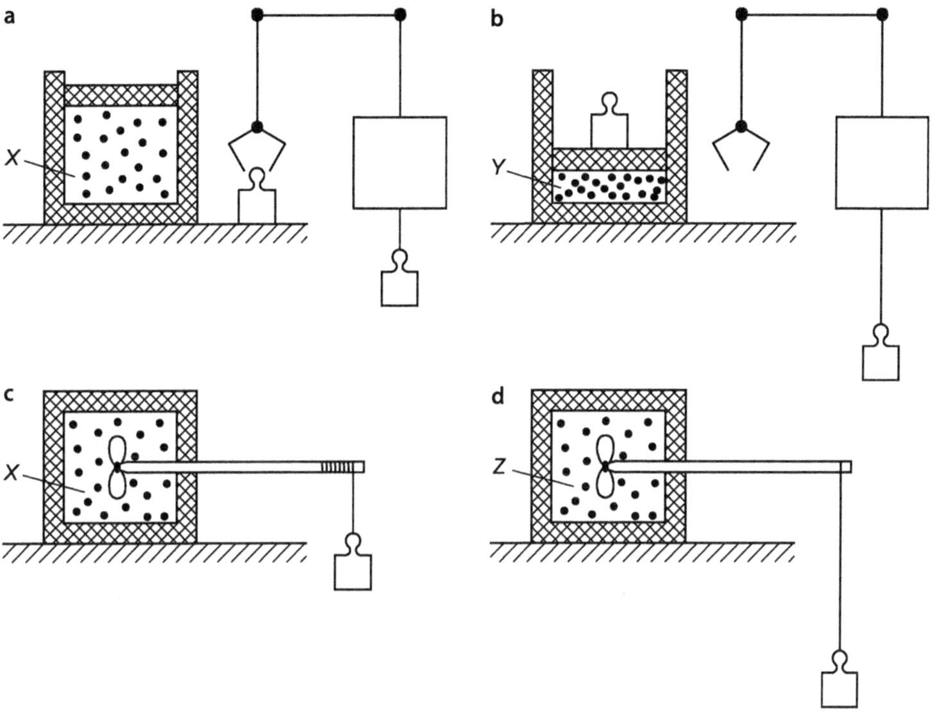

Fig. 2.2. *Adiabatic accessibility for a gas.* Equilibrium states of a gas in a well insulated container before (a, c) and after (b, d) the system has interacted with a device and a weight. The details of the construction of the device in (a) and (b) are unimportant. Strictly speaking, the device must include a little lever which initiates the lifting of the weight (from (a) to (b)) or the lowering of the weight (from (b) to (a)). In (c) and (d), an electrical heater driven by a generator could be used instead of a fan

energy from a weight which can change its position in the gravity field. At the end of the process, the temperature of the gas will have increased. In the case of a perfectly insulated container, the gas will remain at this elevated temperature once and for all.

Our device consisting of the container, the piston and the weight-lifting apparatus does not return to its initial state after the considered process has been carried out. However, we *can* in principle return the device into its initial state in a purely mechanical way. Since this is the case, and only the position of the weight has changed, the considered process fulfills the conditions for adiabatic accessibility. We therefore conclude that $X \prec Y$.

Is it possible to fully reverse the process? In general, this is not the case. This is due to the friction between the piston and the container as well as the viscosity of the moving gas. When the gas is quickly compressed, part of the mechanical work performed by the moving piston is converted into a turbulent flow of the gas. After the piston has come to rest, the turbulence will decay and the kinetic energy of the moving gas will be dissipated. Since it is impossible to reverse this turbulent dissipation of flow energy (Nobody has ever observed that a quiescent fluid spontaneously cools down and starts to move!), a full reversal of the compression is not possible. The same is true for the effect of mechanical friction (A piston never spontaneously cools down in order to start moving).

However, if we were able to eliminate any friction effect from the system and if we carried out the compression so slowly that any flow of the gas could be avoided (except for the unidirectional flow perpendicular to the face of the moving piston), then we could approach an idealized situation in which $X \succ Y$ holds in addition to $X \prec Y$. In this idealized case, we would have $X \overset{A}{\sim} Y$. If we refined our experiments by using six small weights instead of one large weight, we would come to the conclusion that not only the initial and final states are adiabatically equivalent but the intermediate states are adiabatically equivalent to X and Y as well.

Next we investigate the system shown in Figs. 2.2c and d. Let us assume that the same amount of gas as in the previous example is enclosed in a container with constant volume. Let us further assume that a fan is installed in the container which can be driven by a weight which moves in the gravitational field of the Earth. Let us finally assume that the state in Fig. 2.2c is the same as in Fig. 2.2a. Now consider a process in which we start from the state in Fig. 2.2c. We let the fan rotate for some time, stop the rotation, wait for any internal flow to decay and arrive at state Z shown in Fig. 2.2d. The volumes in states Z and X are obviously equal. The internal energy has increased by the same amount by which the potential energy of the weight has decreased. After the process, the apparatus is in the same state as in the beginning. (The mass of the rope is assumed to be negligible.) Therefore we have $X \prec Z$. For the same reasons as discussed in connection with Fig. 2.1a, it is not possible to adiabatically reverse the considered process. Indeed, it has never been observed that a gas cools spontaneously and sets a fan into motion which in turn raises a weight. We can therefore conclude that $Z \nprec X$ and $X \prec\prec Z$.

Let us now summarize the results of our thought experiments in a similar way as in our introductory example about financial accessibility displayed in Fig. 1.1. Our summary is shown in Fig. 2.3 which represents the thermodynamic analog to Fig. 1.1. A trivial property of the table in Fig. 2.3a is that all three elements on the diagonal of the matrix are **YES**. This information has no deep physical meaning and reflects the obvious fact that each state is adiabatically accessible from itself. The remaining four nontrivial entries, three times **YES** and one **NO**, are the result of our hypothetical experiments.

For a complete description of the outcome of all possible experiments, two statements about the mutual adiabatic accessibility of Y and Z are still missing and would

Fig. 2.3.
From adiabatic accessibility to entropy. Adiabatic accessibility (a) and entropy (b) of the equilibrium states shown in Fig. 2.2. The numerical values of S_0 and S_1 are unimportant for understanding. They only have to satisfy the condition $S_0 < S_1$

	X	Y	Z
X	Y	Y	Y
Y	Y	Y	
Z	N		Y

a

$S(X) = S_0$
$S(Y) = S_0$
$S(Z) = S_1$

b

require additional experimental work. If we included the intermediate states of our refined thought experiment involving the small weights, we would still have to perform a total of $8 \times 8 = 64$ experiments to completely characterize the adiabatic accessibility of all our states. Fortunately, however, this is unnecessary for the same reason as it was not necessary in Sect. 1.2 to determine all 64 relations of financial accessibility in the German fairy tale *Hans in Luck*. In the introductory example, the introduction of the concept of the value of an object has allowed us to reduce the amount of information from 64 to 8. Now we ask a similar question, namely whether we could not encode the equilibrium states of Fig. 2.3a using a certain quantity. If we could find such a quantity, we could determine the adiabatic accessibility of two equilibrium states of thermodynamic systems by just comparing the values of this quantity between the two states in question.

Finding the existence of such a quantity is one of the most important scientific achievements of the nineteenth century. This quantity was introduced by Rudolf Clausius in 1865 and called entropy.

To introduce our new quantity for the system considered in Fig. 2.2, let us assume that each of the considered three states is characterized by a quantity S, denoted entropy, which satisfies the following two conditions. First, if two states are adiabatically equivalent, their entropies should be equal. Second, if two states are not adiabatically equivalent, the state which is not adiabatically accessible from the other should have the lower entropy. (Notice that $X \prec\prec Y$ implies a *decreasing* value for financial accessibility and an *increasing* entropy for adiabatic accessibility if X is the initial and Y the final state.)

Since the states X and Y in Fig. 2.2a are adiabatically equivalent, we assign the same entropy $S(X) = S(Y) = S_0$ to both of them, as shown in Fig. 2.3b. For the time being, we will neither care about the exact numerical value nor the unit of S_0. We have further observed that $X \prec\prec Z$. Hence we assign a value S_1 to X which has the property $S_1 > S_0$ but is otherwise arbitrary. Without having to perform a single additional experiment, we can immediately complete the table in Fig. 2.3a. Indeed, by comparing the entropies of X and Y we find $S(Y) < S(Z)$ from which we readily conclude $Y \prec\prec Z$. Hence we can write **Y** in line 3, column 4 and **N** in line 4, column 3. If we had included the six additional states into our consideration, which are adiabatically equivalent and have the entropy S_0, we could immediately complete all 64 fields of the table that corresponds to the one shown in Fig. 2.3a. This would have saved us as much as fifty experiments! It is remarkable that the mere existence of entropy entails such a tremendous simplification.

With this preview of entropy done, we finish our discussion of the physical content of adiabatic accessibility. For the time being, we shall delete the terms *temperature*, *heat*, *hot* and *cold* from our vocabulary. We remind the reader that the central conclusion of this chapter is that the equilibrium states of the thermodynamic system shown in Fig. 2.2 can be sorted using the order relation $\prec$ and that this order relation can be encoded in a quantity called entropy. Our next goal is to answer the question whether it is possible to define entropy for arbitrary systems and whether there are universal properties which our order relation $\prec$ must satisfy so that an entropy function $S(X)$ exists.

2.5 Coordinates

The attentive reader will have noticed that up to now we have not asked ourselves how the equilibrium states X, Y, and Z can be mathematically described. We have deliberately deferred the discussion of this question in order to demonstrate that the existence of entropy is independent of the way in which we parameterize our state space (Remember that the state space, loosely speaking, refers to all possible equilibrium states of a thermodynamic system). However, if we are to perform thermodynamic computations for real-life problems, it is indispensible to endow our state space with *coordinates*. In classical mechanics the time-dependent position of a mass point is uniquely described by the Cartesian coordinates $\boldsymbol{x}(t) = [x(t), y(t), z(t)]$ of the Euclidian space. In thermodynamics the definition of coordinates requires more subtle considerations.

The fundamental coordinate of thermodynamics is the internal energy U. Its existence is based on our experience and is one of the fundamentals of thermodynamics. For our present discussion, the mere existence of this quantity is sufficient. For a detailed discussion of its physical nature, we refer the reader to textbooks on thermodynamics, for instance Moran and Shapiro (1995). The existence of the energy is postulated by the first law of thermodynamics. Following Lieb and Yngvason (1999), we formulate this law in the following way:

> **First law of thermodynamics** (physical formulation). The amount of work W done by the outside world in going adiabatically from one state of the system to another is independent of the manner in which this transition is carried out.

This formulation goes back to Max Planck (Planck 1964) and differs from many textbook definitions in that it does not contain the concept of heat. Planck had recognized that heat is difficult to define without entropy. He therefore referred to "the mechanical equivalent of all forms of energy that are supplied to the thermodynamic system" – an expression which is equivalent to the present formulation of the first law. As a consequence, the quantity W can be interpreted as the change $U_2 - U_1 = W$ of a variable U which is called internal energy. From a mathematical point of view, the content of the first law can be expressed very succinctly as:

> **First law of thermodynamics** (mathematical formulation). U is a coordinate.

The first law has played an important role in the history of thermodynamics. But in the framework of the Lieb-Yngvason theory, its content reduces to the definition of a special coordinate U, the internal energy, which we shall call more generally *energy coordinate*.

The building block of thermodynamics is a substance with variable volume V, as for instance one kilogram of water vapor. As shown in Fig. 2.4a, its state $X = (U, V)$ is represented by the two coordinates internal energy and volume and can therefore be considered as a point in a two-dimensional space. A system with only one energy coordinate, like the present one, is called a *simple system*. The volume is referred to as a *work coordinate*. Figures 2.4b, c and d illustrate how important the correct choice of

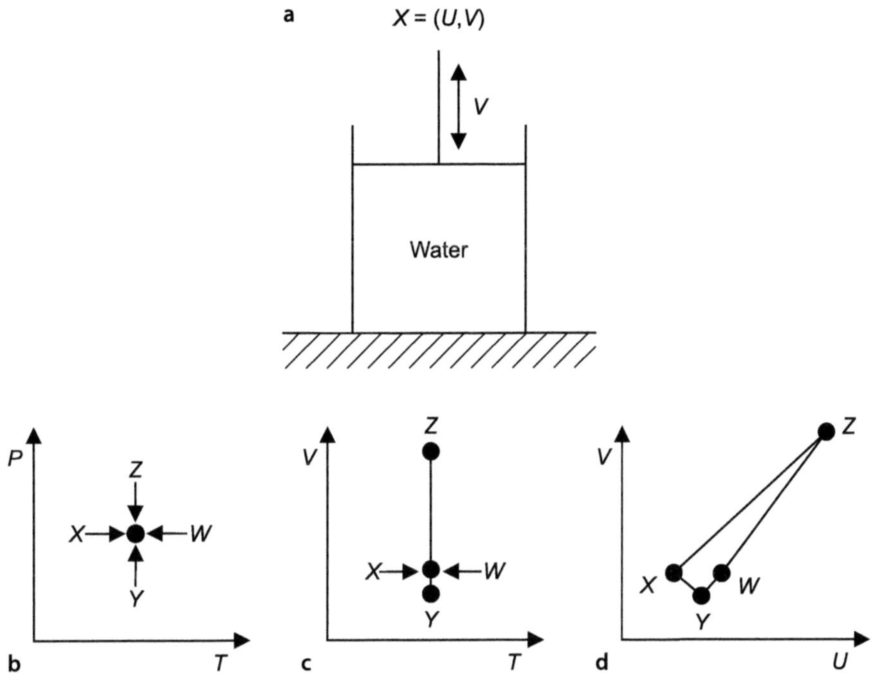

Fig. 2.4. *Simple systems – the building blocks of thermodynamics.* (a) Simple system with a two-dimensional state space characterized by the internal energy U and volume V. Representations of the triple point of water in the T-p-plane (b), T-V-plane (c), and U-V-plane (d) show that only the coordinates $X = (U, V)$ permit an unambiguous description of the state of the system. The states X (ice), Y (liquid water), Z (water vapor) and W (mixture of liquid water and water vapor with the same volume as the ice) are indistinguishable in the T-p-plane. In the T-V-plane, X and W are indistinguishable. Only in the U-V-plane can the states X and W be distinguished. The coordinates of the states for 1 kg of water are $X = (-333.40\ \text{kJ}, 1.0908 \times 10^{-3}\ \text{m}^3)$, $Y = (0, 1.0002 \times 10^{-3}\ \text{m}^3)$, $Z = (2\,375.3\ \text{kJ}, 206.136\ \text{m}^3)$, and $W = (1.05\ \text{J}, 1.0908 \times 10^{-3}\ \text{m}^3)$. The state W comprises approximately 0.44 μg water vapor and (1 kg – 0.44 μg) of liquid water. The triangle in (d) is not to scale

coordinates is for a unique characterization of the state of a system. Figure 2.4b shows that the seemingly natural coordinates for, namely temperature and pressure, are not suited to uniquely describe the state of water at the triple point. At $T = 0.01$ °C and $p = 0.006113$ bar, water can be a solid, a liquid or a gas as well as a combination of any of these three states. If we chose $X = (T, p)$ as the coordinates of our state space, these mixture states would be indistinguishable. If we had taken temperature and volume as coordinates instead, the situation would have improved only slightly. As Fig. 2.4c shows, the triple point would have become a triple line, but the coordinates of the state X (pure ice) and W (mixture) could still not be distinguished. Only if we use energy and volume as coordinates, the triple line becomes a triple surface and we can distinguish X and W, see Fig. 2.4d.

The system shown in Fig. 2.4 contains two degenerate cases which deserve particular attention. The first such case pertains to a system whose work coordinate does not change. We call such a system a *reservoir* or a *thermometer*. The state space $X = (U)$ of such a system is one-dimensional and contains only the internal energy. The second degenerate case is a *mechanical system*. In a mechanical system, all states are adiabatically equivalent. For instance, an isothermal mechanical spring undergoing slow (quasi-static) processes is characterized by the work coordinate x and its potential energy $kx^2/2$ where k denotes the stiffness of the spring. The thermodynamic state of the spring is therefore described by the coordinates $X = (kx^2/2, x)$. This expression shows that the energy and work coordinates of a mechanical system are uniquely related to each other. The mechanical system is degenerate in the sense that all states have the same entropy.

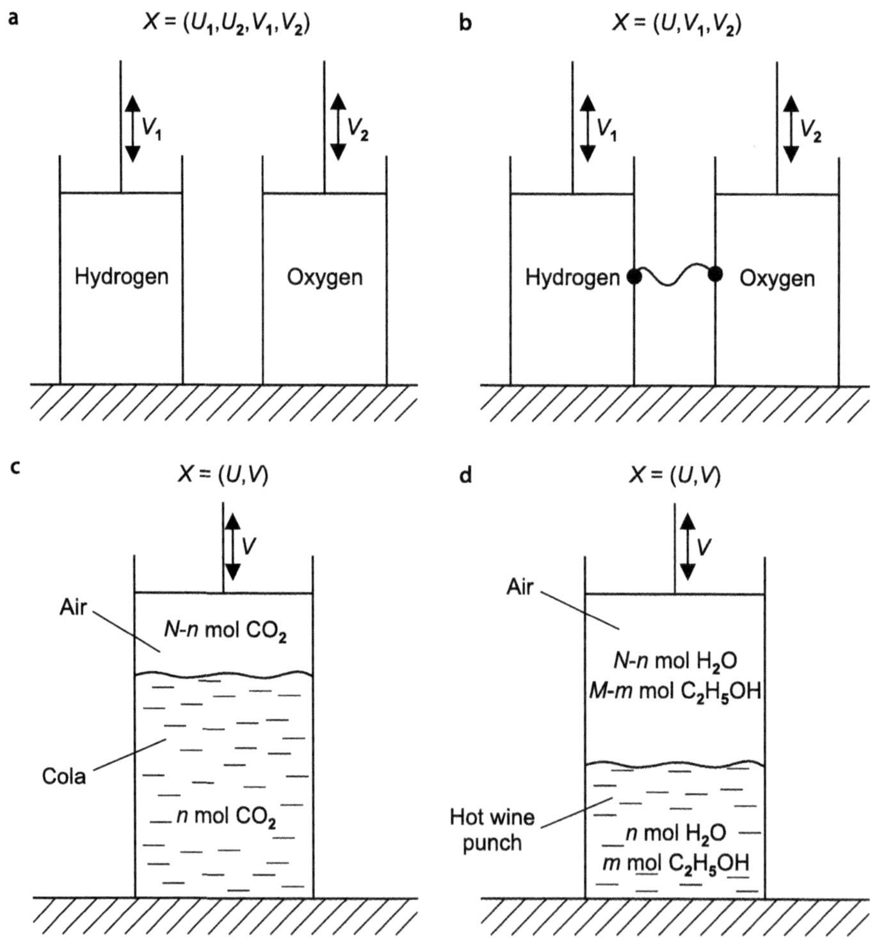

Fig. 2.5. *Coordinates of selected systems.* (a) Compound system with two energy coordinates and two work coordinates, (b) simple system which represents the thermal join of the two subsystems in (a). This system has two work coordinates. The distribution of the internal energies among the two subsystems is described by the variable α which is mentioned in the text. (c) Simple system with a single work coordinate. The system can be characterized by an additional variable which describes the distribution of a dissolved gas between two phases. (d) Simple system with a single work coordinate. The system can be characterized by two additional variables describing the distribution of water and alcohol between the two phases

Figure 2.5 shows the coordinates of other thermodynamic systems. If we consider two non-interacting thermodynamic systems as a unity, we obtain a *compound system*. Mathematically speaking, the state space of a compound system is the product of the state spaces of the individual systems. If we connect the two subsystems, using a copper wire as in Fig. 2.5b for instance, their temperatures become equal after a sufficiently long time. The internal energies U_1 and U_2 of the two subsystems can no longer be varied independently. Hence the compound system with formerly four degrees of freedom has lost one degree of freedom. It has become a simple system with the coordinates U, V_1, and V_2. The distribution of the internal energy among the subsystems can be written in the form $U_1 = \alpha U$, $U_2 = (1 - \alpha)U$ using the parameter α with $0 \leq \alpha \leq 1$. Observe that α is neither a thermodynamic coordinate nor a degree of freedom. However, this quantity will become useful when we discuss the definition of temperature in Sect. 4.2. We wish to emphasize that we currently use the word temperature only as a "mnemonic guide" (Lieb and Yngvason 1999) to characterize a process of energy exchange between two thermodynamic systems. We could equally well imagine that each system is painted with a temperature-sensitive color. Our observation that "the temperatures of both systems become equal" would then translate into "the colors of both systems become equal".

The systems shown in Figs. 2.5c and 2.5d display a property which is similar to a feature of the example from Fig. 2.5b and which is encountered in many practical applications. They contain quantities such as for instance the amount of carbon dioxide dissolved in a beverage or the amount of alcohol that has evaporated from a cup of mulled wine. These quantities cannot be controlled from the outside. They rather attain certain values once the system has settled in an equilibrium state. They do not therefore represent coordinates in the sense of our previous discussion. However, it is often the case that such quantities are of interest and need to be determined. It is therefore often necessary to regard them formally as thermodynamic variables and to include them into the state space. Such systems play an important role in chemical engineering and in the food industry. They also demonstrate that a "simple" system in the sense of our mathematical definition is not necessarily simple in the sense of chemistry. For instance, coffee contained in a Thermos® bottle (including the air above the free surface of the coffee) is a simple thermodynamic system in the sense of our definition. It has a two-dimensional state space consisting of energy and volume. However, coffee contains about 200 substances. Each of these substances is dissolved in the liquid phase to a large extent and is contained in the air to a smaller extent. The parameters describing the distributions of each of these substances between the liquid and gaseous phases can be formally included into a phase space of the form $X = (U, V, n_1, n_2, \ldots, n_{200})$. Here n_1 is the amount of substance 1 in water, $N_1 - n_1$ the amount of the same substance in air and N_1 the total amount of the substance in moles. This extended 202-dimensional phase space still represents a "simple" system according to our terminology.

The definition of coordinates enables us to represent the results of our thought experiments from Figs. 2.2 and 2.3 in the U-V-space. The result is shown in Fig. 2.6. The states X and Y, as well as the intermediate states discussed earlier, are all adiabatically equivalent. Therefore they are characterized by the same value of the entropy S_0. We can thus regard them as being located at the same height S_0 on the "entropy hill"

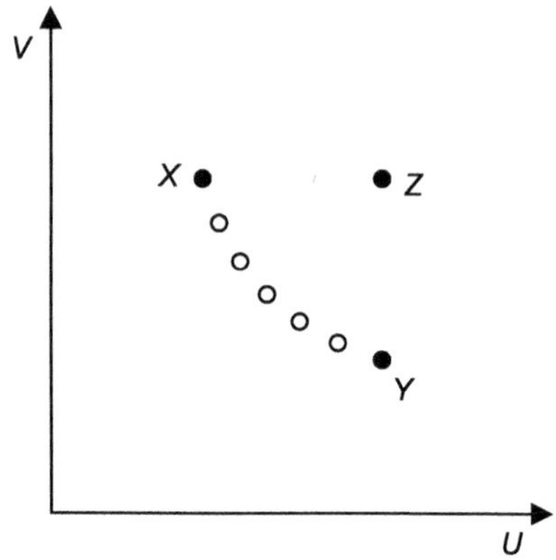

Fig. 2.6.
Equilibrium states in the state space. Representation of the equilibrium states X, Y, and Z (defined in Fig. 2.2) in the two-dimensional state space consisting of internal energy and volume. Additionally, empty circles mark states which were adiabatically accessible from X and Y if several small weights were used instead of a single large one

$S(U, V)$. The hill grows in northeastern direction. At the point Z its height S_1 is greater than S_0. The determination of the exact shape of this entropy hill is one of the central tasks of thermodynamics.

2.6 Properties of Adiabatic Accessibility

The mathematical formulation of any theory in physics proceeds in two steps. Both of them are based on observations and experimental evidence. The first step is the definition of the basic mathematical objects of the theory. The second step consists of the formulation of rules which connect the basic objects. Once the objects and the rules have been defined, the theory can be used to make predictions.

Lieb and Yngvason's formulation of thermodynamics differs from the traditional one in that the number of basic mathematical objects, also called *primitive concepts*, is reduced to the absolute minimum. Moreover, the rules are formulated with the greatest possible amount of logic and mathematical rigor. As a consequence of this mathematical accuracy, the rules may appear formal and sometimes impenetrable to the beginner. Readers who are not interested in the details of the mathematical foundations should therefore skip the following sections and proceed directly to Chap. 3.

The primitive concepts of the Lieb-Yngvason theory are *equilibrium states*, *coordinates* and *adiabatic accessibility*. In what follows, we wish to use our experience from everyday life in order to formulate the basic rules of thermodynamics. This goal does not sound very concrete at the moment. However, the goal will become much more obvious if we remember how classical mechanics are formulated. The primitive concepts of classical mechanics are the *mass* m of a point as well as its *trajectory* $\mathbf{r}(t)$ in the three-dimensional space as a function of time t under the influence of force $F(t)$. The rule connecting these primitive concepts is Newton's second law which states that *mass times acceleration is equal to force*. Its mathematical form is an ordinary differential equation and reads

$$m\frac{\mathrm{d}^2\mathbf{r}}{\mathrm{d}t^2} = \mathbf{F} \tag{2.1}$$

Our goal is to formulate the thermodynamic counterpart of this equation. In contrast to mechanics, this shall require some additional preparatory mathematical work. But we should not hesitate to deal with new mathematics. Newton and Leibniz also had to invent calculus in order to establish the laws of mechanics.

The primitive concepts and rules of thermodynamics have been formulated as recently as 1999 by the physicists Elliott Lieb and Jakob Yngvason (Lieb and Yngvason 1999). They have formulated fifteen axioms which the order relation $\prec$ must satisfy. Based on these axioms, they have derived the *entropy principle* as the central law of thermodynamics which we will extensively discuss in Chap. 3. We will discuss some of the fifteen axioms, but not all. We shall see that in spite of their abstract nature, they are deeply rooted in our everyday experience. Rather than explaining all axioms in the following text, we list them in Appendix B in order to motivate the reader to study Lieb and Yngvason's original work.

The first rule that the order relation $\prec$ has to satisfy is *reflexivity* and has the form $X \overset{A}{\sim} X$. Reflexivity means that each state is adiabatically accessible from itself. This requirement looks trivial but is indispensible for the logical completeness of the theory. Moreover, adiabatic accessibility has to satisfy a number of other conditions of which *comparability*, *transitivity*, *consistency*, *stability* and *convex combination* are particularly important. We shall explain them in more detail now.

A – Comparability

Figure 2.7 shows the results of two fictitious experiments in which the adiabatic accessibility of states A, B, C etc. of a thermodynamic system have been tested. Figures 2.7a and b differ from each other in one aspect. In Fig. 2.7a there are pairs of states, for instance A and B, for which neither $A \prec B$ nor $A \succ B$ holds. We say that such states are

	A	B	C	D	E	F	G	H	I	K
A	Y	N	Y	Y	Y	Y	Y	Y	Y	Y
B	N	Y	N	N	N	N	N	N	N	Y
C	N	N	Y	N	N	N	N	N	N	N
D	N	N	N	Y	N	N	N	N	N	N
E	Y	N	N	N	Y	N	N	N	N	N
F	N	N	N	N	N	Y	N	N	N	N
G	N	N	N	N	N	N	Y	N	N	N
H	N	N	N	N	N	N	N	Y	N	N
I	N	N	N	N	N	N	N	N	Y	N
K	N	N	N	N	N	N	N	N	N	Y

a

	A	B	C	D	E	F	G	H	I	K
A	Y	Y	Y	Y	Y	Y	Y	Y	Y	Y
B	N	Y	Y	Y	Y	Y	Y	Y	Y	Y
C	N	N	Y	Y	N	N	N	N	Y	Y
D	N	N	Y	Y	N	N	N	N	Y	Y
E	N	N	Y	Y	Y	Y	N	Y	Y	Y
F	N	N	Y	Y	Y	Y	N	Y	Y	Y
G	N	Y	Y	Y	Y	Y	Y	Y	Y	Y
H	N	N	Y	Y	Y	Y	N	Y	Y	Y
I	N	N	Y	Y	N	N	N	N	Y	Y
K	N	N	Y	Y	N	N	N	N	Y	Y

b

Fig. 2.7. *Comparability of equilibrium states.* Results of two fictitious series of experiments in which the adiabatic accessibility of states has been investigated. Which of the two tables violates the comparison principle and cannot therefore be described by an entropy function?

not comparable. In Fig. 2.7b, by contrast, for each pair of states at least one of the statements $A \prec B$ or $A \succ B$ is true. We will refer to such states as *comparable*. A central requirement is that the adiabatic accessibility fulfills the following principle.

> **Comparison principle.** For any two states X and Y in the same state space, either $X \prec Y$ or $Y \prec X$ holds.

Figure 2.7a violates the comparison principle whereas Fig. 2.7b satisfies it. It is hard to believe that there is any physical content behind such an abstract statement. However, Fig. 2.8 shows us that this is not the case.

Let us ignore thermodynamics for a moment and imagine that the Figs. 2.7 and 2.8 represent the accessibility of downhill skiers located at different positions during a downhill race. We define the *forward sector* of a certain position as the set of all points which can be reached by passive downhill sliding. Figure 2.8 shows us that the forward sectors of the states in Figs. 2.7a and b differ from each other significantly. The forward sectors in Fig. 2.8a, which correspond to the data of Fig. 2.7a, intersect each other, whereas those in Fig. 2.8b are distinct. For the example shown in Fig. 2.8a, a person in position A cannot access position B. From position B, position A is inaccessible as well.

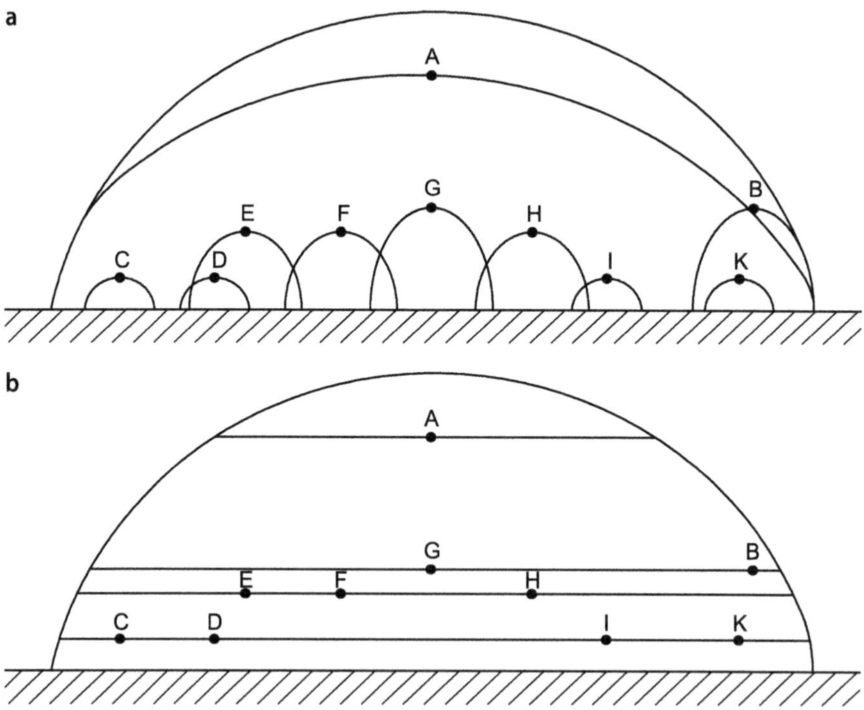

Fig. 2.8. *Comparability and the geometry of forward sectors.* Imagining the equilibrium states of Fig. 2.7 as locations of a skier on a hill, case (a) corresponds to intersecting forward sectors as in Fig. 2.7a whereas case (b) corresponds to non-intersecting forward sectors as in Fig. 2.7b. Hence the thermodynamic analog of case (a) cannot be described by an entropy function

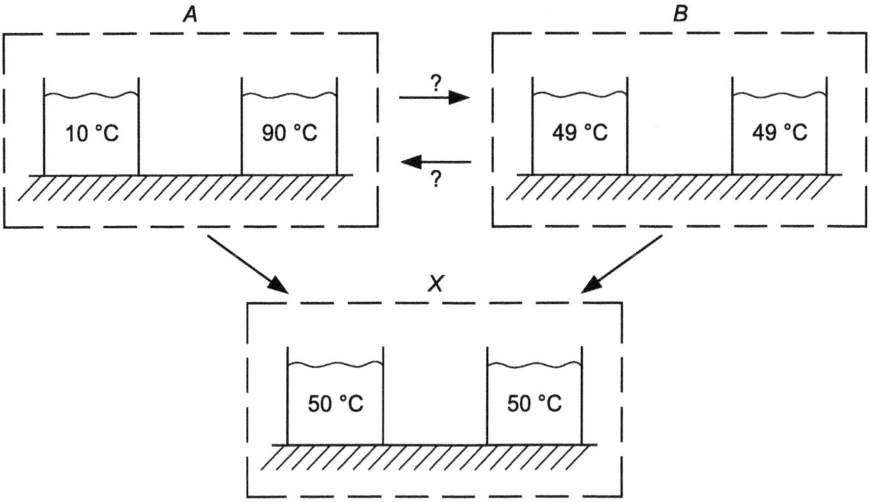

Fig. 2.9. *Comparability and heat engines.* From everyday experience, we know that the forward sectors of the states A and B contain state X. But everyday experience fails if we are to answer the questions "$A \prec B$?" and "$B \prec A$?". The comparison principle requires that at least one of these questions can be answered with **YES**. If $A \prec B$, a heat engine must exist which performs work and reduces the average temperature of the water from 50 °C to 49 °C. If $B \prec A$, a refrigeration system must exist which creates temperature differences

In Fig. 2.8b, B is accessible from A but A is inaccessible from B. The former case corresponds to a skier with friction whereas the latter corresponds to frictionless sliding. This example shows us that a seemingly innocent requirement like comparability can reflect the absence of energy dissipation in an example from classical mechanics. However, we remind the reader that this example has nothing to do with thermodynamics. Let us now return to the question which of our everyday experience with thermodynamic systems is reflected in the comparison principle.

Let A represent the state of a compound system which consists of one kilogram of water with a temperature of 10 °C and one kilogram of water with a temperature of 90 °C, as shown in Fig. 2.9. (Remember that the word "temperature" serves as a mnemonic guide and could be replaced by "voltage of a thermocouple"). Let B be the same system in the state when both subsystems have the same temperature 49 °C. Let both states correspond to the same pressure and assume that the specific heat capacity of water is constant. Are the states A and B comparable? Everyday experience tells us that if we let the two subsystems interact, the initial state A will evolve into the final state X (2 kg of water at 50 °C). Furthermore, it has never been observed that two kilograms of water spontaneously exchange energy and develop a temperature difference. Hence we have $A \prec\prec X$. If our system is initially in state B and we put it into a microwave oven for a moment, it will evolve into state X as well. It is obvious that this process cannot be adiabatically reversed, which implies $B \prec\prec X$. We seem to be in a similar situation as the skier in Fig. 2.8a. We have shown that the forward sectors of both A and B have at least one point in common, namely X. But we do not know whether $A \prec B$ or $A \succ B$ hold. We do not even know if A and B are comparable at all.

The comparison principle requires that at least one of the relations $A \prec B$ or $A \succ B$ be true. This postulate has far-reaching consequences. Indeed, if $A \prec B$ is to be true, the comparison principle requires the existence of an apparatus which brings the system into a final state whose temperature is below the mean temperature of 50 °C. If $A \succ B$ is to hold, the comparison principle enforces the existence of a device which creates a temperature difference of 80 °C in 2 kg of water which were initially at the same temperature. We know that the former device is a heat engine, for instance a Stirling engine, whereas the latter is a refrigeration system like for instance the air conditioning system in a car. It is remarkable that the mere property of comparability implies the existence of such systems, though it does not tell us how to construct such engines.

The application of the comparison principle to systems undergoing mixing and chemical reactions has even more surprising consequences. We illustrate this using the following example.

Consider the states A (1 kg of a stochiometric mixture of hydrogen and oxygen at 20 °C and atmospheric pressure) and B (1 kg of water vapor at 10 °C and atmospheric pressure). Are these states comparable? If we let the mixture react, we obtain water vapor with a temperature above 1 000 °C whose state shall be denoted with X. Since the explosion is obviously an adiabatic process, we can write $A \prec X$. It has never been observed that hot water vapor spontaneously disintegrates into a mixture of hydrogen and oxygen. We can therefore assume that $A \not\succ X$ and therefore $A \prec\prec X$. (A word of caution is in order here. The impossibility of spontaneous disintegration of water vapor into hydrogen and oxygen does not automatically imply $A \not\succ X$. Such an adiabatic process could be in principle possible if it includes raising a weight in order to satisfy the energy balance. We shall nevertheless assume that $A \not\succ X$ is true.) We further know that we can adiabatically transform water vapor from state B to state X by simple heating. The opposite is not true. From this we conclude $B \prec\prec X$.

We are again in a situation now where the forward sectors of A and B have at least one common point, namely X. Again we cannot tell from our everyday experience whether the states A and B are comparable, i.e., if one of the relations $A \prec B$ or $A \succ B$ is true. Strictly speaking, we are not allowed to apply the comparability principle to the present example because A and B belong to different state spaces. A is a member of the state space of hydrogen-oxygen mixtures whereas B belongs to the state space of water. However, Lieb and Yngvason have shown that under some weak assumptions the comparison principle still holds. In the case of mixtures and chemical reactions it is necessary that the systems consist of the same amount of matter. For instance, if the initial state consists of one mole of oxygen and two moles of hydrogen, the final state must consist of one mole of water. The consequences of comparability are surprising, as we shall see next.

If the comparison hypothesis is true for the present example, either $A \prec B$ or $A \succ B$ must hold. In the former case, there must exist a device which lets hydrogen and oxygen react and cool (!) the reaction products adiabatically. In the latter case, there must be an apparatus which adiabatically inverts the exothermal reaction. This putative device would act as an "explosion reversion engine." We do not know whether it is technically possible to build such devices. After having introduced entropy, however, we will be able to compare the entropies $S(A)$ and $S(B)$. On this basis, we shall be able to decide which of these two machines is impossible.

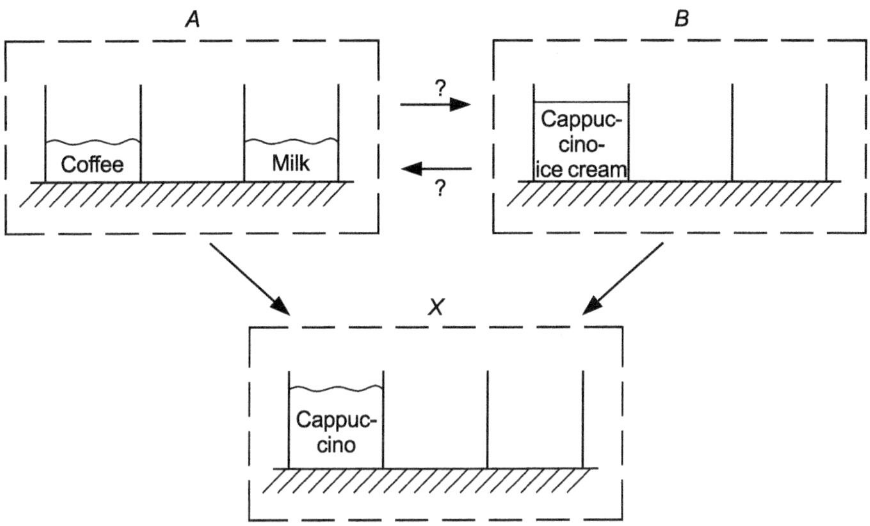

Fig. 2.10. *Comparability and the "cappuccino-ice cream unmixing machine."* From everyday experience we know that the forward sectors of the states A and B contain state X: By mixing coffee with milk (state A) one obtains cappuccino (state X). By melting frozen cappuccino or cappuccino ice cream (state B) one also obtains cappuccino (state X), at least in principle. But our everyday experience fails if we are to answer the questions "$A \prec B$?" and "$B \prec A$?". The comparison principle requires that at least one of these questions can be answered with **YES**. If $A \prec B$, an engine must exist which adiabatically transforms warm coffee and warm milk into cappuccino icecream. By contrast, if $B \prec A$, a device must exist which adiabatically transforms cappuccino icecream into coffee and milk

Another interesting consequence of the comparison principle emerges if we consider mixtures. This case is illustrated in Fig. 2.10, using the example of mixing coffee and milk. Since this example is a straightforward generalization of our previous two examples, we shall not discuss it in detail. A brief explanation is given in the caption for Fig. 2.10.

Summarizing the conclusions that we have drawn from the examples shown in Figs. 2.9 and 2.10, we can state the following. The comparison hypothesis ensures that the forward sectors of equilibrium states of a thermodynamic system do not intersect so that a situation like in Fig. 2.8a is impossible in the state space of a thermodynamic system. We have shown that this fact has important consequences concerning the existence of heat engines and refrigeration systems. Since the comparison principle also holds for systems undergoing mixing and chemical reactions, it also entails the existence of fictitious "explosion reversion devices" and "cappuccino-icecream unmixing machines" (cf. Fig. 2.10). The importance of comparability had already been recognized by Giles (1964). But it was not until 1999 that Lieb and Yngvason have demonstrated that the comparison principle can be derived from even more fundamental postulates which are listed in Appendix B. Furthermore, they could show that, under weak assumptions, the comparison principle also holds for systems undergoing mixing and chemical reactions.

B – Transitivity

Apart from reflexivity and comparability, adiabatic accessibility must be *transitive*. The meaning of transitivity is the following.

Transitivity. $X \prec Y$ and $Y \prec Z$ imply $X \prec Z$.

At first glance, this rule does not seem to contain any physics. However, this is not the case, as the following argument shows. We consider water in the three states X, Y, and Z which correspond to a solid, a liquid and a gas, respectively. The precise coordinates of the states are not important for our discussion. To be specific, however, we could require all three states to have the same pressure. In order to illustrate the consequences of a violation of transitivity we assume for a moment that $X \prec Y$ and $Y \prec Z$ hold, while $X \prec Z$ does not. This would imply, for instance, that one could use a microwave oven to melt ice and to let water boil but it would be impossible to adiabatically transform ice into water vapor. This of course is in contrast to our everyday experience which tells us that we can indubitably transform ice adiabatically into water vapor if we keep the ice in the microwave oven for a sufficiently long time. More generally, $X \not\prec Z$ would violate our experience that one can divide any adiabatic process into several processes and proceed via intermediate states like Y.

The consequences of transitivity in combination with comparability are even more far reaching. If $X \prec Z$ didn't hold, the comparison principle would require that $Z \prec X$ be true. This means that one could transform water vapor adiabatically into ice at constant pressure. Taking into account that such a hypothetic process would have to satisfy conservation of energy, the energy extracted from the vapor and from the liquid water would have to be compensated by raising a weight, in agreement with the definition of adiabatic accessibility. Such a process contradicts our experience. If we could adiabatically extract energy from water vapor, we could also extract energy from the oceans and transform it directly into work. All energy problems would be solved at once. This unfortunately is not the case.

This example shows us that transitivity together with comparability reflect our everyday experience that a thermodynamic process can be split into several parts or that adiabatic processes executed one after another can be regarded as a single adiabatic process.

C – Consistency

Adiabatic accessibility has to satisfy another condition which is called *consistency*. Its definition is as follows.

Consistency. If $X \prec X'$ and $Y \prec Y'$ holds for two systems then the compound system must obey $(X, Y) \prec (X', Y')$.

The situation concerning the physical content of consistency seems to be similar to our previous discussions of comparability and transitivity. However, Fig. 2.11 shows that this is not the case. If the relation $\prec$ were not consistent, this would have strange consequences. Consider two thermodynamic systems. Let us assume that one of them

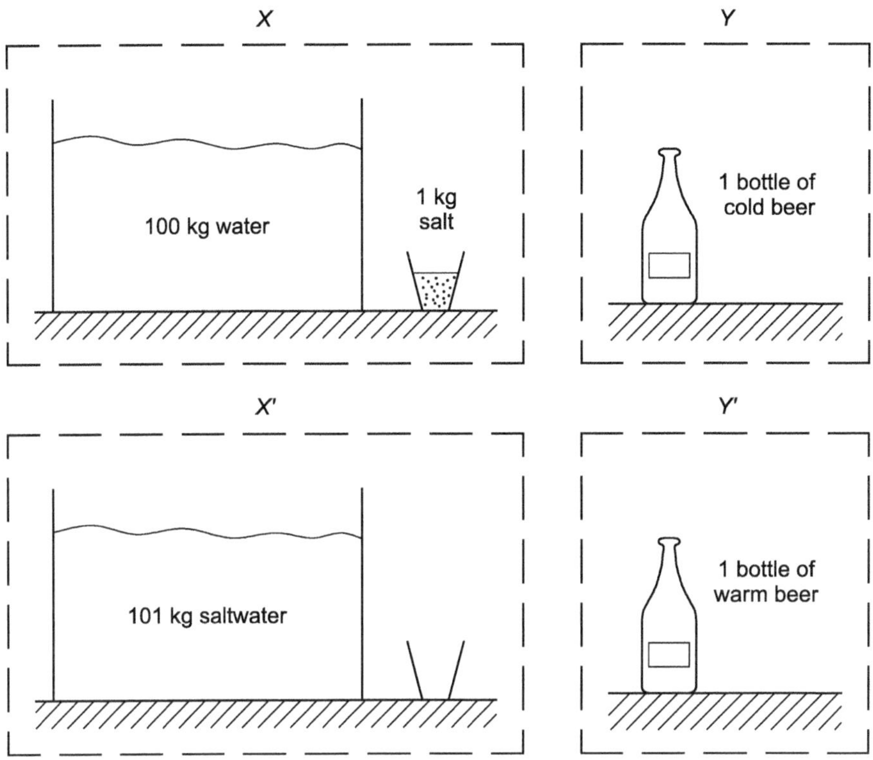

Fig. 2.11. *Consistency.* If adiabatic accessibility were not consistent, one could reverse dissolution of salt in water and heating a bottle of beer by considering the two systems to be a compound system

consists of water and salt whereas the other one consists of cold beer. Let each system independently undergo the adiabatic processes sketched in Fig. 2.12. If consistency would not hold, then the final state would cease to be adiabatically accessible from the initial state as soon as we regard the two simple systems as a compound system. The absurdity of thermodynamics without consistency becomes entirely obvious if we note that by virtue of comparability, $(X, Y) \not\prec (X', Y')$ would imply $(X', Y') \not\prec (X, Y)$. This means that the theoretical unification of salt water and warm beer to a compound system would permit us to adiabatically unmix salt water and simultaneously cool down a quantity of beer. This example violates our everyday experience and moreover illustrates the physical content of consistency.

Closely related to consistency is the possibility of *splitting* and *recombining* thermodynamic systems. This property was already mentioned in Sect. 2.1. Adiabatic accessibility should have the property that splitting of a system into two parts or recombination of two systems with the same thermodynamic properties into one single system are adiabatic processes. Mathematically speaking, this condition reads $X \overset{A}{\sim} (\lambda X, (1-\lambda)X)$ where $0 < \lambda < 1$. Adiabatic accessibility must also obey *scaling invariance.* If $X \prec Y$ holds, then $\lambda X \prec \lambda Y$ must hold as well where λ is an arbitrary nonnegative real number.

D – Stability

Another property of adiabatic accessibility is *stability*. Its mathematical definition is as follows.

Stability. If for some pair of states X and Y the relation $(X, \epsilon Z_0) \prec (Y, \epsilon Z_1)$ holds for a sequence of ϵ tending to zero and some states Z_0 and Z_1 then $X \prec Y$.

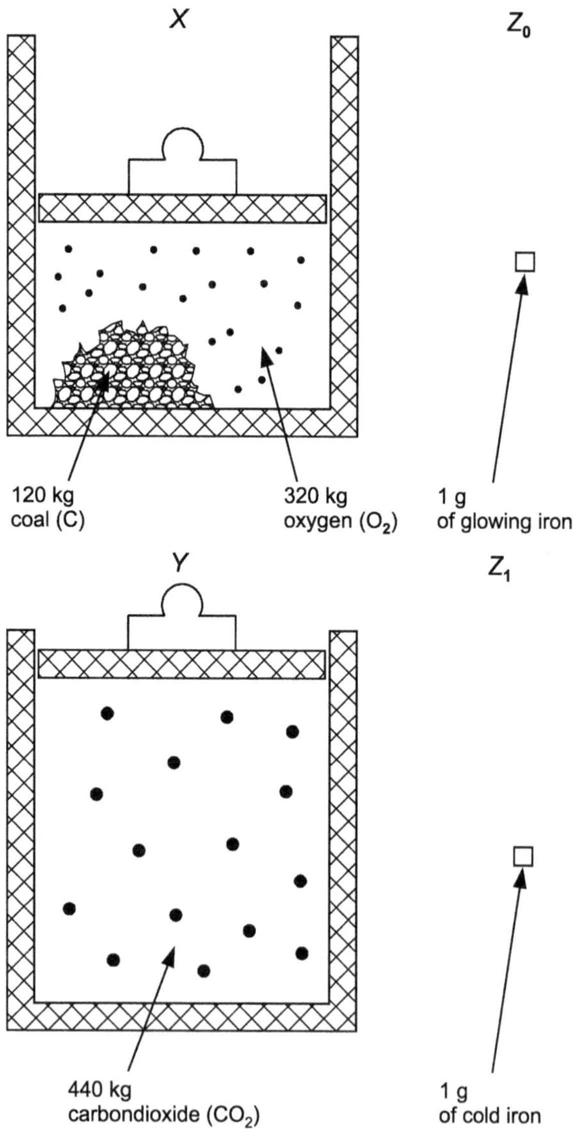

Fig. 2.12.
Stability. If adiabatic accessibility were not stable (in the mathematical sense of axiom A-6 in Appendix B) the combustibility of coal would not be an intrinsic property of the carbon-oxygen system but would depend on the accidental availability of an igniter

The physical meaning of stability is illustrated in Fig. 2.12. We know that coal and air react to carbon dioxide in the presence of a sufficiently hot object, for instance a glowing piece of iron. We should say more accurately that carbon and oxygen react to carbon dioxide but we will stick to the previous terms. The thermodynamic state (X, Z_0) representing coal, air, and glowing iron and the state (Y, Z_1) describing CO_2 and cold iron obviously obey $(X, Z_0) \prec (Y, Z_1)$. We further know that the possibility to burn coal is independent on whether we ignite it using one gram (Z_0), one milligram $(0.001\ Z_0)$ or an arbitrarily small fraction of a gram, which is symbolically written as (ϵZ_0). The mathematical property of stability ensures that the adiabatic accessibility persists in the limit $\epsilon \to 0$. It thereby makes sure that the combustibility of a thermodynamic system is an intrinsic property, or as Lieb and Yngvason expressed it: "stability means simply that one cannot increase the set of accessible states with an infinitesimal grain of dust."

E – Condition of Convex Combination

Finally we wish to explain that adiabatic accessibility has to satisfy the condition of convex combination. Before explaining this condition, we have to introduce the mathematical concept of convex combination.

> **Convex combination.** The state $Z = tX + (1 - t)Y$ with $0 \le t \le 1$ is called the convex combination of the states X and Y.

As illustrated in Fig. 2.13a, the convex combination of two states X and Y can be geometrically interpreted as the set of all points located at the line connecting these points in the state space. For instance, the convex combination of one kilogram of a mixture of ice and water in the state X and one kilogram of a mixture of liquid water and water vapor in the state Y involves all linear combinations $Z = \alpha X + \beta Y$ which correspond to a mixture of ice, liquid water and water vapor. In order to ensure that the convex combination has the same mass as its constituents, it is necessary that $\alpha + \beta = 1$ holds. Figure 2.13d shows the convex combination for the given example in the particular case $t = 1/2$. We can verify by inspection of this figure that the energy and the volume of the convex combination are the arithmetic means of those of the states X and Y. According to the axioms of thermodynamics, listed in Appendix B, the convex combination must satisfy the following condition.

> **Condition of convex combination.** The convex combination $Z = tX + (1 - t)Y$ must be adiabatically accessible from the state $(tX, (1 - t)Y)$, i.e., $(tX, (1 - t)Y) \prec tX + (1 - t)Y$.

The meaning of this condition is illustrated in Fig. 2.13. Figure 2.13c shows a system which is in the state $(X/2, Y/2)$ where X and Y denote the states of one kilogram of water. The brackets indicate that we are dealing with a compound system. This fact is highlighted by the presence of a thermally and mechanically insulating wall. The condition of convex combination tells us that that by removing the wall the convex combination of X and Y is adiabatically accessible from $(X/2, Y/2)$.

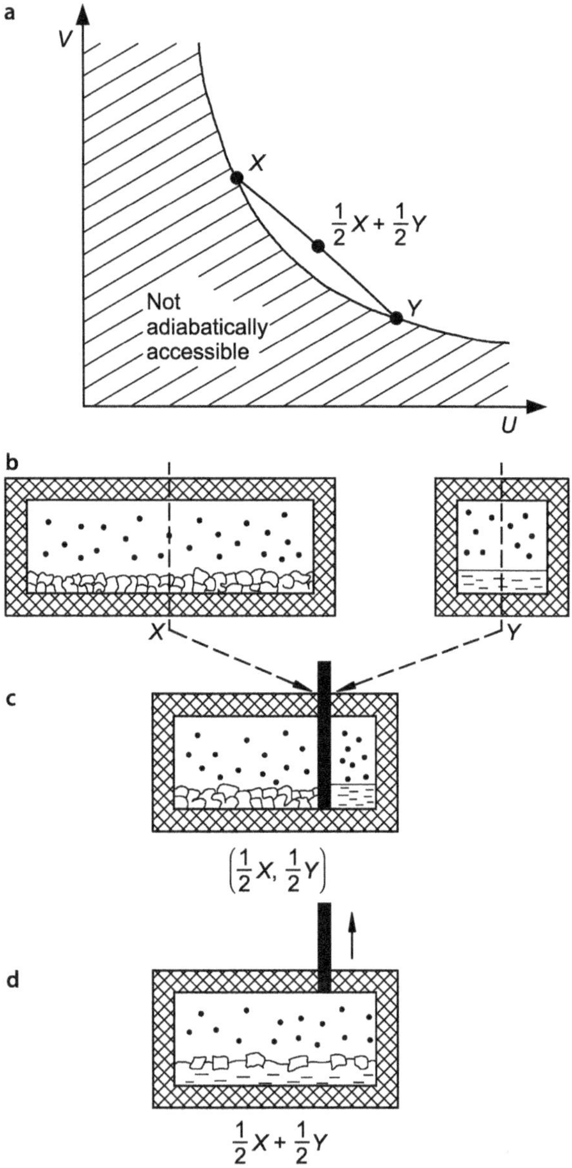

Fig. 2.13.
Convex combination. (a) Illustration of the convex combination of the states X and Y of 1 kg of water shown in (b). X is a mixture of ice and water vapor, whereas Y describes a mixture of liquid water and water vapor. The set of all possible convex combinations is represented by the straight line which connects X and Y in (a). In particular, this set contains the state $X/2 + Y/2$ shown in (d). The property of convex combination (axiom A-7 in Appendix B) states that this state has to be adiabatically accessible from the state shown in (c). (a) is drawn for the particular case when X and Y have the same forward sector which corresponds to the dashed area

Chapter 3 Entropy

Chapter 3

Entropy

In Sect. 2.4 we learned that we can assign entropy values to equilibrium states of thermodynamic systems and that this entropy permits us to make predictions about the adiabatic accessibility. Up to now the entropies S_0 and S_1 had neither a numerical value nor a physical unit. Here we shall show how the entropy can actually be determined on the basis of the properties postulated in Sect. 2.6. For the sake of simplicity, we will restrict ourselves to an explanation of the basic ideas and refer those readers who are interested in the mathematical details to the original paper (Lieb and Yngvason 1999). This work is of particular interest to readers who wish to convince themselves that the properties listed in Sect. 2.6 are indeed all necessary for the proof of the main theorem – the entropy principle.

3.1 Entropy of Water

Before developing a procedure for the determination of entropy, let us remember how we have determined the value W of an object X in our introductory example of Sect. 1.2. The definition $W(X) = \min\{\lambda : \lambda Y_0 \prec X\}$ involved two components, namely a *scale* and a *market*. More precisely, we have defined a scale by specifying a unit Y_0 (e.g. 1 kg of gold) and assigning a value (e.g., 10 000 €) to it. We then entered a fictitious market and performed a series of attempts to financially access X from λY_0 (e.g. $\lambda = 0.3$). The minimum of all possible λ with the property that X is accessible from λY_0 then gave the dimensionless value $W = \lambda$. In order to obtain the dimensional value, for instance in Euros, we had to multiply this value with our unit, for instance $0.3 \times 10\,000$ € $= 3\,000$ €.

To determine entropy we need two similar components: an *entropy scale* and a *Lieb-Yngvason machine*.

Figure 3.1 shows how we can define the entropy scale which we need to determine the entropy of a sample in state X. In order to be specific, let us assume that the sample whose entropy we wish to determine consists of one kilogram of liquid water. To define the entropy scale we need to specify two *reference states* X_0 and X_1 which can be reproducibly realized in an experiment.

Let X_0 denote the state of one kilogram of water at the melting point under atmospheric pressure ($p = 1.013$ bar $= 1.013 \times 10^5$ N m^{-2}). This is not a uniquely defined system yet. Indeed, water at the melting point can consist of either 10% ice and 90% liquid water or 90% ice and 10% liquid water or of any other combination of complementary

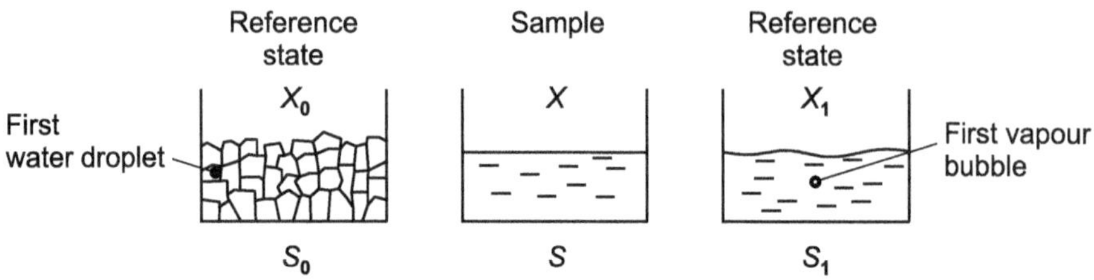

Fig. 3.1. *Definition of an entropy scale.* To determine the entropy of a system in state X, an entropy scale is necessary. This scale is defined by two reference states X_0 and X_1 which obey the property $X_0 \prec\prec X_1$. If X is the state of one kilogram of water, one could for instance use $X_0 =$ (1 kg of ice at the melting point at atmospheric pressure) and $X_1 =$ (1 kg of liquid water at the boiling point at atmospheric pressure)

quantities of solid and liquid water. In order to make the reference state X_0 unique, we assume that it consists of 100% ice and 0% liquid water. In order to illustrate that such a system begins to melt as soon as its internal energy experiences the slightest increase, we have added a tiny drop of water in Fig. 3.1. To simplify the notation we refer to this reference state simply as "ice."

Let the second reference state X_1 represent one kilogram of water at the boiling point, again at atmospheric pressure. As with the first reference state, this is not a uniquely defined state yet, because our system can consist of any mixture of liquid water and water vapor. In order to make X_1 unique, we stipulate that it consists of 100% liquid water and 0% water vapor. We symbolically add a tiny vapor bubble in order to highlight that the smallest increase of the internal energy will let the system boil. For simplicity, we call the reference state X_1 "hot water."

Each of the so-defined reference states $X_0 = (U_0, V_0)$ and $X_1 = (U_1, V_1)$ represents a point in the two-dimensional state space of one kilogram of water. In general, the reference states have to satisfy the condition $X_0 \prec\prec X_1$ but can otherwise be arbitrary. We assign them the entropies $S(X_0) = S_0$ and $S(X_1) = S_1$ with $S_0 < S_1$. This defines our entropy scale. For the time being we regard entropy as a non-dimensional quantity and set $S_0 = 0$ and $S_1 = 1$. Later we will define an entropy unit. The property $S_0 < S_1$ has the consequence that entropy increases when we slide from one thermodynamic state to the other in a series of adiabatic processes. By contrast, the property $W_0 > W_1$ of our introductory example implies that the value decreases when exchanging a piece of gold for a goose. Uncoordinated action produces entropy and destroys value!

To determine the entropy of a state X (under the restriction $X_0 \prec X \prec X_1$ which will be released later) we have to compare our sample with the entropy unit. To perform such a comparison we have to divide the unit into small subunits in a similar way as we would engrave millimeters into a yardstick. We must then compare our sample with different entropy subunits until we have determined the equivalence of our sample with a certain number of subunits.

To accomplish this task, we define a compound system which consists of $(1 - \lambda)$ kilograms of ice in state X_0 and λ kilograms of hot water in state X_1. We assume that the parameter λ is in the interval $0 \leq \lambda \leq 1$. Our entropy yardstick is a one-parameter family of states and has the form

$$((1 - \lambda)X_0, \lambda X_1) \tag{3.1}$$

The comparison between the sample and the entropy scale is accomplished by experimentally analyzing for which values of λ the state of the sample X is adiabatically accessible from $((1 - \lambda)X_0, \lambda X_1)$.

Such experimental tests are tedious in practice. For this reason, we shall assume the existence of a fictitious device which embodies all existing relations of adiabatic accessibility for all thermodynamic systems. This device is the thermodynamic analogue of the market in our introductory example. In honor of the scientists who developed the present formulation of thermodynamics we call such a device a *Lieb-Yngvason machine.*

A Lieb-Yngvason machine is shown in Fig. 3.2. It is a device containing two slots and two lights. The left light is marked with **Y** for **YES** whereas the right light is marked with **N** for **NO**. In order to analyze whether a state Y is adiabatically accessible from a state X, one has to insert a thermodynamic system in state X into the left slot. The thermodynamic system in the state Y is inserted into the right slot. Notice that the systems inserted into the slots have to consist of the same amount of matter. Once the systems have been inserted, the Lieb-Yngvason machine performs the following task: If $X \prec Y$ is true, the left light will flash. If $X \prec Y$ is wrong, the right light will flash. The Lieb-Yngvason machine embodies all knowledge about thermodynamics that has been accumulated by mankind in the past and that will be added in future. It is a symbolic representation of all experimental and computational devices which can be used to analyze the adiabatic accessibility of thermodynamic states. For the definition of entropy, it is not important whether it is actually possible to build such a device. The mere assumption if its existence is crucial. However, it is important that the Lieb-Yngvason machine satisfies all axioms discussed in Sect. 2.6 and listed in Appendix B, and the left and right slots contain samples with identical material composition. Let us summarize:

Lieb-Yngvason machine. A Lieb-Yngvason machine is a fictitious device that satisfies all axioms of adiabatic accessibility given in Appendix B and determines whether for two arbitrary states X and Y of the same thermodynamic system the relation $X \prec Y$ holds or not.

Let us now determine the entropy of a sample using a Lieb-Yngvason machine. This shall be done by comparing the sample with the entropy scale defined by Eq. 3.1. Remember that the family of systems described by Eq. 3.1 consists of the same amount of matter as the sample and the members of this family differ from each other only by the fraction λ of hot water. The comparison principle ensures that such a comparison is always possible.

The procedure for the determination of entropy using the Lieb-Yngvason machine boils down to answering the following question: What is the lowest amount of "expensive" ice and the complementary amount of "cheap" hot water from which the state X can be adiabatically accessed? We use the terms expensive and cheap in order to highlight that it is easy to transform ice into hot water but difficult to turn hot water into ice. Moreover, these terms relate the present case to the intuitive understanding that we have acquired in the introductory example of Sect. 1.2.

In a first trial, illustrated in Fig. 3.2a, we let the Lieb-Yngvason machine analyze whether X is accessible from the member of entropy scale having $\lambda = 0$. To do this, we place a system with entropy $S = 0$ into the left and the sample into the right slot of the Lieb-Yngvason machine. The system with $\lambda = 0$ consists of one kilogram of ice and no

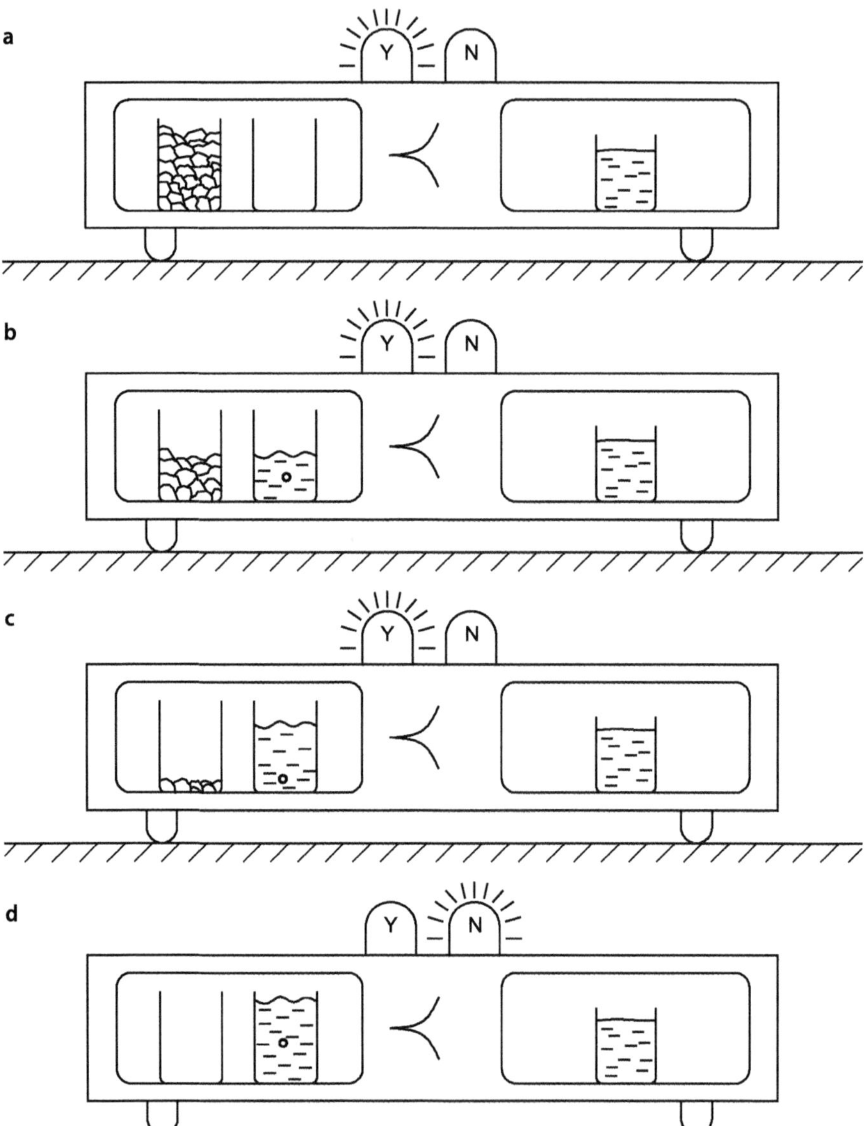

Fig. 3.2. *Lieb-Yngvason machine and the entropy of a simple system.* Method for determining the entropy of one kilogram of water using a Lieb-Yngvason machine. The machine answers each question about the adiabatic accessibility of a given state from another one either with **YES** or **NO**. The machine compares a compound system with known entropy standing in the left slot with a given system with unknown entropy standing in the right slot. The compound system consists of $1-\lambda$ kg of ice and λ kg of water at the boiling point. The examples are shown in the order of decreasing amounts of ice and increasing amounts of hot water according to (a) $\lambda_a = 0$, (b) $\lambda_b > 0$, (c) $\lambda_c > \lambda_b$ and (d) $\lambda_d = 1$. The maximum value $\lambda = \mu$ for which $((1-\lambda)X_0, \lambda X_1) \prec X$ must be in the interval $\lambda_c \leq \mu \leq \lambda_d$. By refining the step width, one could determine μ with any desired accuracy

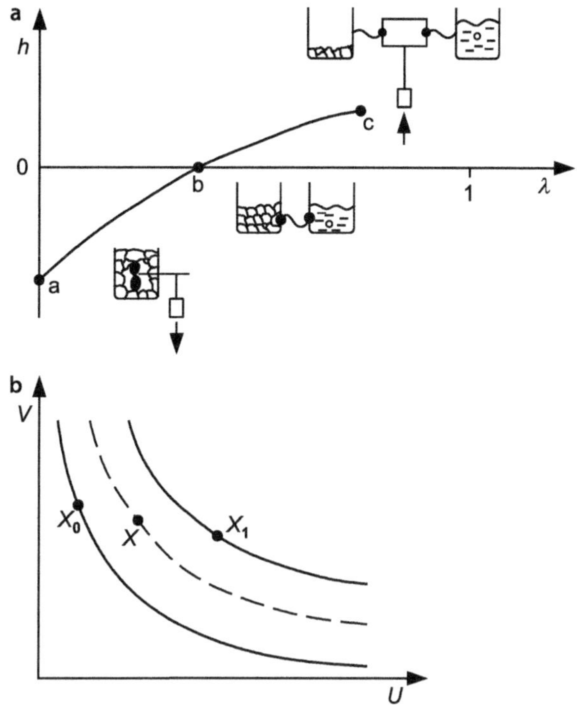

Fig. 3.3.
Adiabatic processes and isentropes. (a) Change of height h of a weight after an adiabatic process which transforms a thermodynamic system from state $((1-\lambda)X_0, \lambda X_1)$ into state X. Points a, b and c correspond to adiabatic processes which transform the systems contained in the left opening of a Lieb-Yngvason machine (shown in Fig. 3.2a–c) into the states shown in the right opening, respectively. a: Stirring, b: heat conduction, c: lifting device consisting of a thermocouple and an electric motor. When the ice melts, its volume decreases slightly. The work which the surrounding air performs on the system has been neglected when drawing the curve $h(\lambda)$. Otherwise, the curve would have to be slightly below the curve actually drawn, except in point c. (b) Representation of the reference states X_0 and X_1, as well as of the state X whose entropy is to be determined. Also shown are the lines of constant entropy in the state space which are also called isentropes

hot water. Once the Lieb-Yngvason machine will have accomplished the analysis, the left light will flash. This indicates that $((1-\lambda)X_0, \lambda X_1) \prec X$ for $\lambda = 0$. This result can be readily understood because we can turn ice into the state of the sample by sufficiently long mechanical stirring. The required work is symbolically represented in Fig. 3.3a as a change in height h of a weight. Since stirring requires work which has to be obtained by lowering a weight, $h < 0$ for $\lambda = 0$. At this point, we should remember that we need to find the maximum of λ for which the relation $((1-\lambda)X_0, \lambda X_1) \prec X$ is true. For maximizing λ, the variant shown in Fig. 3.2a is not optimal because it makes no use of the hot water. Moreover, we need to perform mechanical work. Can we adiabatically access X with less ice and more hot water?

Let us now continue our analysis by comparing our sample with systems whose value of λ increases stepwise. Several examples of systems with increasing λ, which is equivalent to increasing entropy, are shown in Figs. 3.2b, c and d. Figure 3.2b demonstrates that X is adiabatically accessible from $((1-\lambda)X_0, \lambda X_1)$ not only for $\lambda = 0$ but also for $\lambda > 0$. (We assume that X and X_0 are not adiabatically equivalent, otherwise we would trivially have $S(X) = 0$ and nothing would remain to be determined). Indeed, if we connect the vessels containing ice and hot water using a copper wire and keep the pressure constant, the specific energies (energy per unit mass) and the specific volumes (volume per unit mass) will approach after a sufficiently long time. If the value of λ has been chosen correctly, the former ice and the initially hot water will be in the states $((1-\lambda)U, (1-\lambda)V)$ and $(\lambda U, \lambda V)$, respectively. By virtue of the properties explained in Sect. 2.6 and Appendix B, we can recombine the two systems into a single system in state $X = (U, V)$ which is identical to the state of the sample. As sketched in Fig. 3.3, this process does not require work ($h = 0$). We have thus shown that X can also be adiabatically accessed from a system with $\lambda > 0$ and without any mechanical work. However, we are not yet done. We now wish to know whether we can access X from a system with an even larger amount of hot water, for instance from the system shown in Fig. 3.2c.

Figure 3.3a shows that this is indeed the case. For instance, we can install a thermocouple between the ice and the hot water. We could use the produced electric energy to raise a weight using an electric motor. We could as well use a Stirling engine provided that it only interacts with the ice and hot water and leaves no other changes in its environment. In this case, there would be $h > 0$, i.e., we would have performed work.

At some point, the potential to increase λ will be exhausted, as Fig. 3.2d illustrates. For some λ, the Lieb-Yngvason machine would respond **NO** to the question whether $((1-\lambda)X_0, \lambda X_1) \prec X$ holds. (We assume that X is not adiabatically equivalent to X_1 otherwise we would trivially have $S(X) = 1$ and there were nothing to prove). If we refined our experiments by decreasing the step size of λ, we would arrive at the result that there is a maximum amount of hot water that we can transform into the state X using a complementary amount of ice. We define this value of λ as the entropy of state X. Our definition of entropy thus takes the form

$$S(X) = \max\{\lambda : ((1-\lambda)X_0, \lambda X_1) \prec X\} \tag{3.2}$$

We denote the so-defined maximum of λ by μ. Then the relation $((1-\mu)X_0, \mu X_1) \stackrel{A}{\sim} X$ holds so that we can also write Eq. 3.2 in the form

$$S(X) = \{\mu : ((1-\mu)X_0, \mu X_1) \stackrel{A}{=} X\} \tag{3.3}$$

With Eq. 3.2 we have accomplished the goal to find a formula which determines the entropy of a simple system in state X. In words Eq. 3.2 can be formulated as follows.

> **Entropy of water.** The entropy of one kilogram of water in state X is determined by the maximum amount of hot water in the state X_1 which can be adiabatically transformed into X using a complementary amount of ice in state X_0.

Figure 3.3b shows that the states of a simple system with one work coordinate which have the same entropy form contour lines in the two-dimensional state space (U, V). At this moment, the entropy defined by Eq. 3.2 is a non-dimensional quantity. In order to assign a unit to the entropy, we relax our preliminary assumption $S(X_0) = 0$ and $S(X_1) = 1$ and define the entropies for our reference states as $S(X_0) = S_0$ and $S(X_1) = S_1$. Then we have

$$S(X) = (1-\mu)S_0 + \mu S_1 \tag{3.4}$$

We now define the entropy unit *Clausius* which we abbreviate as Cl. We assign the values

$$S_0 = -291.6\ \text{Cl} \tag{3.5}$$

$$S_1 = +312.2\ \text{Cl} \tag{3.6}$$

to our reference states. In doing so, we have now uniquely defined the entropy of one kilogram of water as

$$S(X) = -291.6\ \mathrm{Cl} + \mu \times 603.8\ \mathrm{Cl} \tag{3.7}$$

Cl is an outdated entropy unit which is not used anymore and which does not comply with the SI system. We have nevertheless taken the unlawful step to use this unit in order to highlight that *entropy*, not temperature, is the primary quantity of thermodynamics. We will explain in Chap. 4 that *Clausius* is related to the SI units Joule (J) and Kelvin (K) by the relation $1\ \mathrm{Cl} = 4.1868\ \mathrm{J\,K^{-1}}$. It is customary in engineering thermodynamics (see for instance Moran and Shapiro 1995, table A-2, line 1, column 10) to set the entropy of liquid water equal to zero at the triple point (although strictly speaking $S = 0$ should hold at zero temperature according to the so-called third law of thermodynamics). This leads to numerical values of the entropies of the reference states which are not easy to memorize because they are computed on the basis of real thermodynamic property data from the tables A-2 and A-6 of the textbook by Moran and Shapiro (1995).

In defining entropy according to Eq. 3.2, we have implicitly assumed that $X_0 \prec X \prec X_1$ which restricts the set of states for which entropy can be determined. However, our entropy definition can be extended to states with $X \prec\prec X_0$ and $X_1 \prec\prec X$ if we interpret the statement $((1-\lambda)X_0, \lambda X_1) \prec X$ in a generalized way and admit $\lambda < 0$ or $\lambda > 1$. For instance, for $\lambda = -1/2$ the expression $(1.5X_0, -0.5X_1) \prec X$ has to be interpreted as $1.5X_0 \prec (X, 0.5X_1)$ which enables us to extend the definition of entropy to states whose entropy is smaller than -291.6 Cl or larger than $+312.2$ Cl.

3.2 Entropy of Further Substances

Using the result of the previous section, we can write the entropy of water in state X in the generalized form

$$S_{H_2O}(X) = (1-\mu_X)S_0^{H_2O} + \mu_X S_1^{H_2O} \tag{3.8}$$

where we have defined the constants as $S_0^{H_2O} = -291.6$ Cl and $S_1^{H_2O} = 312.2$ Cl. We can extend this definition to other substances. For instance, if we denote the reference states of solid hydrogen at melting temperature, liquid hydrogen at boiling point, solid oxygen at melting temperature and liquid oxygen at boiling point as Y_0, Y_1, Z_0 and Z_1, respectively, we can write the entropy of the states Y of hydrogen and Z of oxygen as

$$S_{H_2}(Y) = (1-\mu_Y)S_0^{H_2} + \mu_Y S_1^{H_2} \tag{3.9}$$

$$S_{O_2}(Z) = (1-\mu_Z)S_0^{O_2} + \mu_Z S_1^{O_2} \tag{3.10}$$

The quantities μ_Y and μ_Z have to be determined using the Lieb-Yngvason machine in the same way as we discussed for water in Fig. 3.2.

Here we encounter a fundamental question. Are we free to choose the entropy constants $S_0^{H_2} = S(Y_0), S_1^{H_2} = S(Y_1), S_0^{O_2} = S(Z_0)$ and $S_1^{O_2} = S(Z_1)$ in the same way as for water?

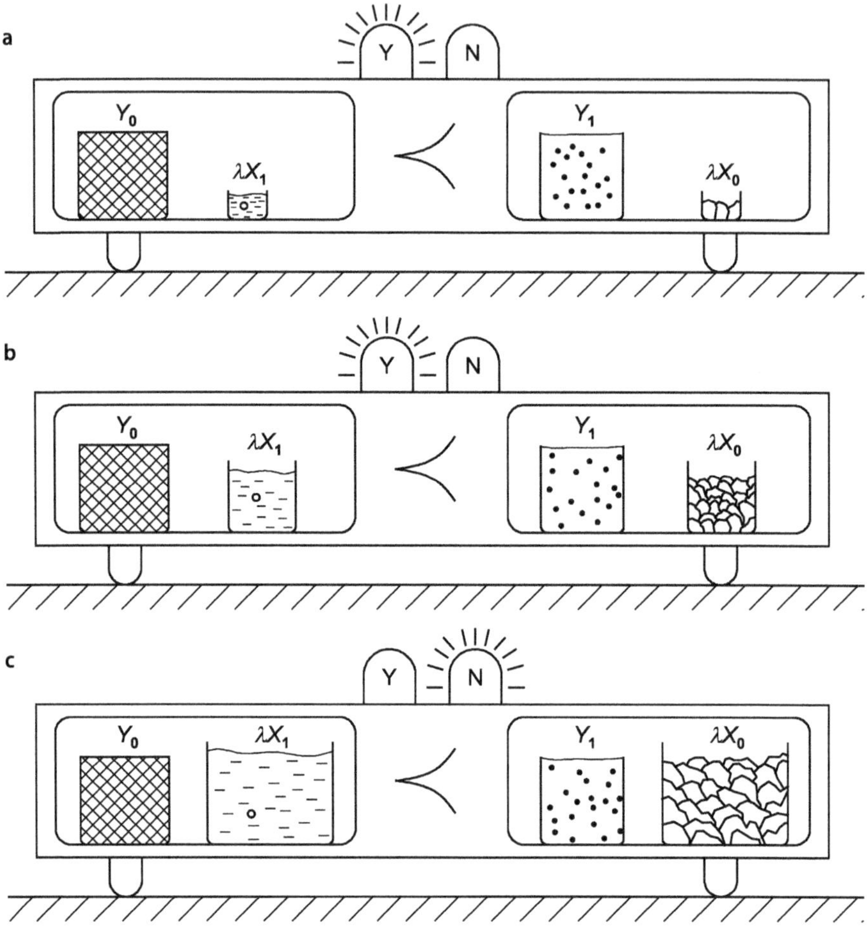

Fig. 3.4. *Calibration of the entropy of hydrogen.* Schematic representation of the procedure for the determination of the entropy difference $S_1^{H_2} - S_0^{H_2}$ of one kilogram of hydrogen in the states Y_0 and Y_1. To accomplish this task, one determines the quantity λ of hot water (in state X_1) which can be adiabatically transformed into ice (in state X_0) when the hydrogen is simultaneously transformed from state Y_0 (solid hydrogen at the melting point) to state Y_1 (liquid hydrogen at the boiling point) for which the entropy is sought. If the maximum of λ is denoted as μ, the entropy difference can be expressed as $S_1^{H_2} - S_0^{H_2} = \mu(S_1^{H_2O} - S_0^{H_2O})$

This is not the case, as we shall show next. Once we have specified the entropy scale for one substance, for instance water, as $S_1^{H_2O} - S_0^{H_2O} = 603.8$ Cl, the scales are fixed for all other substances as well. According to Lieb and Yngvason, we call the determination of these scales *calibration.*

The procedure for calibrating the entropy of hydrogen by comparison with water is sketched in Fig. 3.4. In order to determine by how much the entropy of hydrogen *increases* when we transform it from Y_0 to Y_1, we make use of our knowledge that the entropy of one kilogram of water *decreases* by 603.8 Cl if it undergoes a transition from X_1 to X_0. Consequently, λ kilograms of water lose $\lambda \times 603.8$ Cl of entropy if they undergo the same transition. Roughly speaking, we can "extract" a desired amount of entropy from the water and "inject" it into the hydrogen in order to measure by how much the entropy of the latter system increases.

To accomplish this task, we consider a one-parameter family of compound systems consisting of one kilogram of hydrogen and λ kilograms of water. We wish to determine the maximum amount of hot water which we can adiabatically transform into ice by making use of the hydrogen undergoing a transition from Y_0 to Y_1. It is obvious that we can transform the hydrogen from Y_0 to Y_1 without any water ($\lambda = 0$) by mechanical stirring. But this information is not sufficient because we are interested in the maximum of λ. As shown in Fig. 3.4a, we now use the Lieb-Yngvason machine to determine whether we can also accomplish the transition from Y_0 to Y_1 with $\lambda > 0$. To be specific, let us assume that $\lambda = 0.1$, which corresponds to 100 grams of water. The affirmative answer of the Lieb-Yngvason machine shows that this is possible. Intuitively speaking, we have just found that hydrogen increases its entropy by at least 0.1×603.8 Cl $= 60.38$ Cl when going from Y_0 to Y_1. We next increase the amount of water, see Fig. 3.4b, for instance to $\lambda = 0.5$. Let us assume that the machine responds with **YES** which shows us that the entropy increase of the hydrogen must at least be equal to 0.5×603.8 Cl $= 301.9$ Cl. If we increase λ further, for instance to $\lambda = 2$ as shown in Fig. 3.4c, we will arrive at the point where $(Y_0, \lambda X_1) \nprec (Y_1, \lambda X_0)$. The maximum of λ obviously determines the desired entropy difference in the form

$$S_1^{H2} - S_0^{H2} = \max\{\lambda : \{(Y_0,\lambda X_1) \prec (Y_1,\lambda X_0)\} \tag{3.11}$$

If we denote the maximum of λ by μ as in the case of pure water, $(Y_0, \lambda X_1) \stackrel{A}{\sim} (Y_1, \lambda X_0)$ holds and we can write this equation as

$$S_1^{H2} - S_0^{H2} = \mu \cdot \left(S_1^{H_2O} - S_0^{H_2O}\right) \tag{3.12}$$

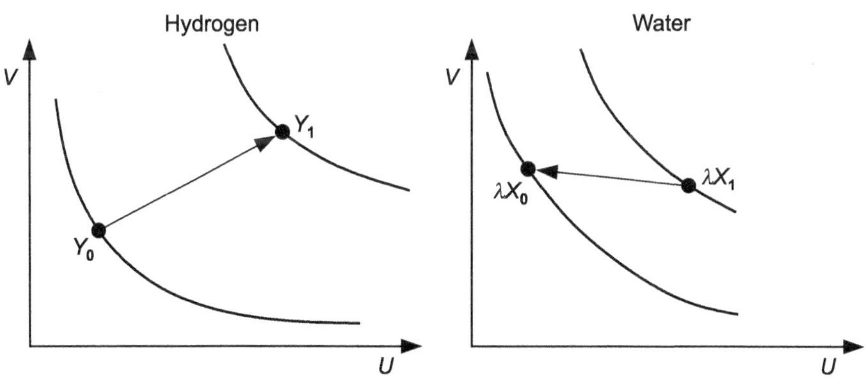

Fig. 3.5. *Calibration procedure in the state space.* Initial and final state of a compound system consisting of hydrogen and water. The initial state corresponds to the systems located in the left slot of the Lieb-Yngvason machine shown in Fig. 3.4a and b. The final state corresponds to the systems located in the right slot of the Lieb-Yngvason machine shown in Fig. 3.4a and b. Notice that during the transformation, the system is not necessarily in a state of thermodynamic equilibrium. Hence the arrows do not necessarily correspond to the actual path of the system in the state space

Figure 3.5 schematically shows the transition in the state space. The same considerations as for hydrogen hold for any other substance as well.

As a result of our thought experiments, we come to the conclusion that for each substance σ only one of the two entropy constants S_0^σ, S_1^σ is a free parameter. If we had for instance chosen S_0^σ, calibration against the fundamental substance water would provide us with a unique value of S_1^σ given by

$$S_1^\sigma = S_0^\sigma + \mu \cdot \left(S_1^{H_2O} - S_0^{H_2O} \right) \tag{3.13}$$

One question still remains. Are we free to choose the remaining constants S_0^σ?

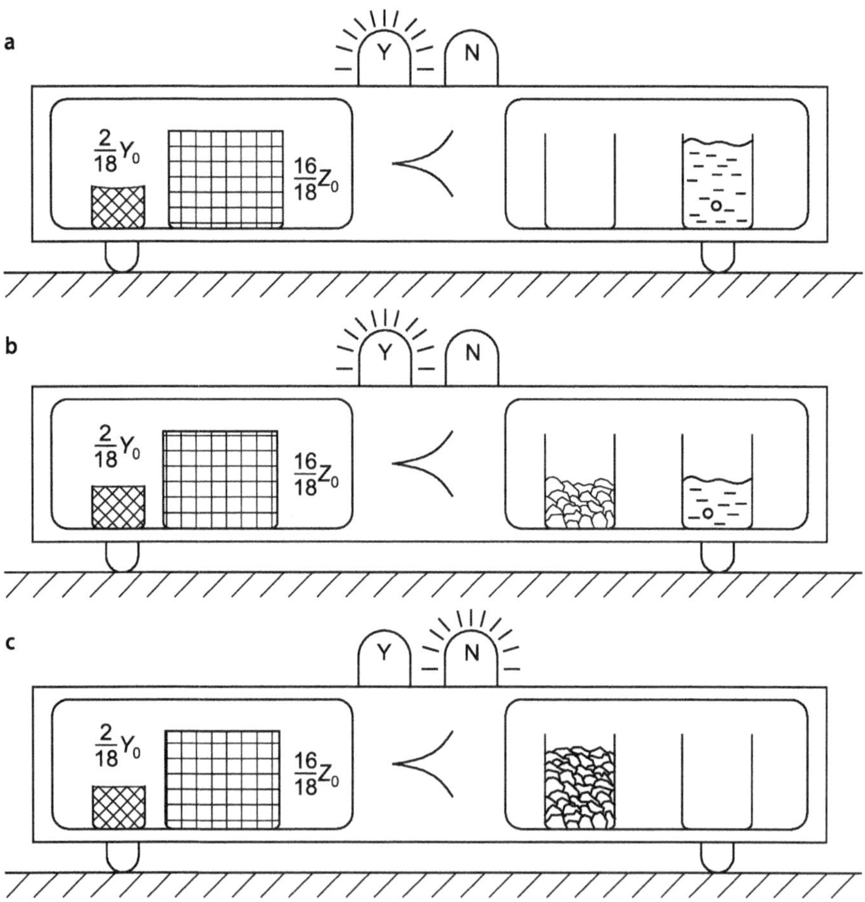

Fig. 3.6. *Entropy and chemical reactions.* Determination of the entropy of an unmixed system consisting of solid hydrogen and solid oxygen shown in the left slot of the Lieb-Yngvason machine. The entropy is determined by comparing this system to a system consisting of melting ice and hot water as shown in the right slot of the Lieb-Yngvason machine. The systems are arranged in the order of decreasing λ. (a) $\lambda = 1$, (b) $0 < \lambda < 1$, (c) $\lambda = 0$. The minimum of λ for which the lamp **Y** shines determines one of the two calibration constants of hydrogen or oxygen

3.3 Mixing and Chemical Reactions

As long as systems do not mix and undergo chemical reactions, the remaining constants S_0^σ are arbitrary. However, mixing and chemical reactions impose constraints upon these constants which we wish to discuss next.

If we transform 2/18 kilograms of hydrogen and 16/18 kilograms of oxygen from the solid states Y_0 and Z_0 into gaseous states and let them mix and explode, we obtain one kilogram of water vapor. The entropy of the initial state is

$$\frac{2}{18} S_0^{H_2} + \frac{16}{18} S_0^{O_2} \tag{3.14}$$

Since this compound system consists of the same amount of matter as one kilogram of water, we can compare it with another compound system that consists of different parts of ice and hot water, as shown in Figs. 3.6 and 3.7. The flashing left lamp in Fig. 3.6a shows us that one kilogram of hot water ($\lambda = 1$, right slot) has a higher entropy than the system in the left slot. (Similar to Sect. 3.1, we exclude the case $(2Y_0/18, 16Z_0/18) \stackrel{A}{\sim} (X_1)$ because this would imply $2S_0^{H_2}/18 + 16S_0^{O_2}/18 = S_1$ and there would be nothing to determine). We decrease λ in small steps as sketched in Figs. 3.6b and c and find a lower bound λ_{min} which we call μ. This lower bound obeys $((2/18)Y_0, (16/18)Z_0) \prec ((1-\mu)X_0, \mu X_1)$ from which we infer that

$$\frac{2}{18} S_0^{H_2} + \frac{16}{18} S_0^{O_2} = (1-\mu) S_0^{H_2O} + \mu S_1^{H_2O} \tag{3.15}$$

This equation shows us that the constants $S_0^{H_2}$ and $S_0^{O_2}$ are not arbitrary but must satisfy one additional condition. As shown in the work of Lieb and Yngvason (1999), the entropy constants of all elements from the periodic table can be expressed by those of one single system, for instance water. With this step done, we can now proceed to formulating our most important theorem, the entropy principle.

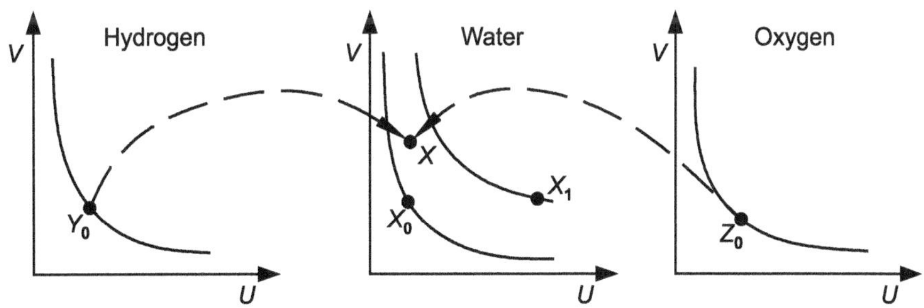

Fig. 3.7. *Reaction of hydrogen and oxygen in state space.* State space representation of the initial and final states of a hydrogen-oxygen system when it undergoes the chemical reaction from the initial state shown in the left part of Fig. 3.6 to a final state consisting of water. Observe that the system is not in thermodynamic equilibrium during the reaction. Moreover, the initial and final states are not in the same state spaces. Hence the arrows do not correspond to the actual path of the system in the state space

3.4 The Entropy Principle

In Sect. 3.1, 3.2 and 3.3 we have shown that to each equilibrium state X of a simple thermodynamic system with one work coordinate we can assign an entropy S. We have also shown that S is unique up to an arbitrary constant. However, our discussion up to now is incomplete for three reasons. First, our explanations are qualitative so far. Our derivation of Eq. 3.2 was not a mathematical proof but merely a result of hand-waving arguments. In order to prove Eq. 3.2, it is necessary to start from the properties of adiabatic accessibility embodied in the axioms from Appendix B, which we have not done. Second, we have restricted our discussion to the simplest possible thermodynamic systems, namely to those with a two-dimensional state space (U, V). For the definition of entropy to be universal, it is indispensible to extend the analysis to arbitrary thermodynamic systems. Finally, we have not elucidated the mathematical and physical properties of the entropy yet. The theorem to be introduced below fills these three gaps.

Based on the complete set of axioms listed in Appendix B, Lieb and Yngvason (1999) were able to prove a theorem which they called the *entropy principle*. The mathematical details of the proof are beyond the scope of the present book, but the final result can be understood without these details. The entropy principle embodies the second law of thermodynamics and reflects its content in a mathematically accurate, logically consistent and physically complete way. The entropy principle has the following form.

Entropy principle. For all equilibrium states X of all thermodynamic systems there exists a real function S called entropy. The entropy is

- **Monotonic:** $X \prec\prec Y \Longrightarrow S(X) < S(Y)$, $X \overset{A}{\sim} Y \Longrightarrow S(X) = S(Y)$;
- **Additive:** $S((X, Y)) = S(X) + S(Y)$ and extensive: $S(tX) = tS(X)$;
- **Concave:** $S(tX + (1-t)Y) \geq tS(X) + (1-t)S(Y)$.

The monotonicity and additivity of entropy follow from the axioms A-1 to A-6 and the comparison principle CP given in Appendix B. The concavity of entropy follows if axiom A-7 is taken into account additionally.

At first glance the entropy principle looks quite formal and does not appeal to our physical intuition. As it will turn out shortly, however, this impression is superficial and wrong. In essence, the entropy principle tells us that in order to predict whether a given final state of a thermodynamic system is adiabatically accessible from a given initial state, it is sufficient that we compare just two numbers, namely the entropies of these states. Moreover, the entropy principle makes sure that these entropies exist. But why is it important to know whether two states are adiabatically accessible? To understand the reason for the importance of adiabatic accessibility, remember that its definition contains the phrase that a "weight may have changed its position in a gravitational field." The understanding is completed if we realize that, for instance, the generation of electricity in a coal-fired power plant is nothing other then an adiabatic process converting a thermodynamic system from an initial state consisting of coal, air and cooling water to a final state consisting of carbon dioxide and warm cooling water and "changing a weight in a gravitational field" (which is equivalent to the generated electric energy). Hence the entropy principle provides us with a tool to answer the question whether a needed amount of electricity can be produced using a given amount of fuel.

To develop our appreciation for the entropy principle further, let us quantitatively demonstrate its far-reaching consequences by briefly returning to a question already posed in Sect. 2.6. There we have asked whether the state Y (2 kg of water at 49 °C) is accessible from the state X (1 kg of water at 10 °C and 1 kg of water at 90 °C). The entropy principle tells us that in order to answer this question we have to compare the entropies $S(X)$ and $S(Y)$. If we find $S(X) < S(Y)$, then $X \prec\prec Y$. This would imply the existence of a device which accomplishes a heat transfer between two systems with a final temperature (49 °C), which is lower than that for passive heat exchange. The existence of such a system would imply that the difference between the internal energies of a final state with 50 °C and a final state with 49 °C must manifest itself in mechanical work. Hence the property $X \prec\prec Y$ enforces the existence of a heat engine. This conclusion is remarkable since we made no assumption about any machine when deriving the entropy principle.

If a comparison between the entropies of X and Y lead to the opposite result, $S(X) > S(Y)$, we would conclude $Y \prec\prec X$ (which we could write as $X \succ\succ Y$). This property would imply the existence of a refrigeration system which adiabatically transforms two kilograms of water with a temperature of 49 °C into one kilogram of water at 10 °C and one kilogram of water at 90 °C. Notice that our use of the term "temperature" only serves to characterize the states in a way that appeals to physical intuition. Given the fact that we have not yet defined temperature, we could have avoided the use of this term by saying "1.3 Volt at a thermocouple" instead of 90 °C.

The previous discussion can be extended to processes involving mixing and chemical reactions such as the mixing process shown in Fig. 2.10. By similar considerations as the previous ones it follows from the entropy principle that under certain conditions there exist devices which adiabatically unmix mixtures and undo chemical reactions.

The content of the entropy principle can be summarized in popular terms in the following way.

> **Law of entropy increase.** In an adiabatic process, the entropy of a system cannot decrease.

3.5 Properties of Entropy

After having explained that the existence of entropy is a consequence of the axioms given in Appendix B and after having formulated Eq. 3.2 for its determination, we would like to discuss the properties of entropy in more detail.

A – Monotonicity

The property of *monotonicity* reflects the fact that entropy encodes the order relation $\prec$. Indeed, the entropy $S(X)$ is a real number which is assigned to each equilibrium state X. If we wish to know whether $X \prec Y$ holds, we only need to compare the entropies of the two states. If we find $S(X) < S(Y)$, we can conclude that $X \prec\prec Y$ provided that X and Y belong to systems with identical physical composition. If the comparison leads to $S(X) = S(Y)$ then we can conclude that $X \overset{A}{\sim} Y$. A succession of processes with $X \prec\prec Y \prec\prec Z$ is encoded in a series $S(X) < S(Y) < S(Z)$ of monotonically increasing entropy values. The function $S(X)$ drastically reduces the amount of information necessary to describe the adiabatic accessibility of the states of a thermodynamic system.

In order to illustrate this property, which was already mentioned in Sect. 2.3, imagine a thermodynamic system which can attain one thousand different states. If we were to completely characterize the adiabatic accessibility of this system using a table analogous to that shown in Fig. 2.3, we would have to deal with a table with one million entries. The monotonicity of entropy permits us to represent the information content of this table by just one thousand real numbers.

B – Additivity

Additivity may seem to be a trivial property of entropy. But this is not the case. Additivity permits one to derive conclusions about adiabatic processes in compound systems which go far beyond the capabilities of our physical intuition. To be more specific, we mention that the property of additivity entails the possibility to adiabatically separate salt water into water and salt provided that an auxiliary system is available such as a copper block, a container filled with gas or a vessel with ice cubes.

Let X_1 denote the state of 1 001 g of salt water. Let us X_2 denote the state of a compound system consisting of 1 000 g of pure water and one gram of salt (not dissolved in the water). As we will show in Sect. 5.7, the entropy of the sample of salt water is by $\Delta S = 1.239$ Cl larger than the sum of the entropies of the constituents of the compound system in state X_2 (The analysis of the last example on the inside of the cover in Appendix G will show that the entropy difference is actually slightly larger). When separating salt water into water and salt, the entropy of the system changes by

$$S(X_2) - S(X_1) = -\Delta S \tag{3.16}$$

The negative sign on the right-hand side of this equation shows us that the entropy decreases. According to the law of entropy increase, this equation demonstrates that there is no adiabatic process which transforms X_1 into X_2 because such a process would lead to a decrease of entropy.

The situation changes fundamentally if we consider a compound system consisting of salt water and a second system, for instance, of ice cubes. We do not use the ice cubes to dilute the salt water. The ice cubes serve only as an auxiliary thermodynamic system. We could also perform our thought experiment using salt water and a block of copper. When one kilogram of ice melts, its entropy increases by 1 219 Cl. Consequently, the entropy of 1.016 g of water changes by 1.239 Cl when it is transformed from the solid state Y_1 to the liquid state Y_2, i.e.,

$$S(Y_2) - S(Y_1) = +\Delta S \tag{3.17}$$

The entropy change associated with the melting of 1.016 g of ice has the same magnitude and opposite sign as the change of entropy when separating 1 001 g of salt water into its constituents. We can express the entropy of the initial state $Z_1 = (X_1, Y_1)$ of the compound system (salt water and ice cubes) and of the final state $Z_2 = (X_2, Y_2)$ of the compound system (water, salt, molten ice) by using the additivity of entropy in the form $S(Z_1) = S(X_1) + S(Y_1)$ and $S(Z_2) = S(X_2) + S(Y_2)$. The change in the total entropy of the compound system can be obtained by adding Eqs. 3.16 and 3.17. We finally arrive at

$$S(Z_2) - S(Z_1) = 0 \tag{3.18}$$

Obviously, the entropy decrease during desalination of salt water is compensated by the entropy increase due to the melting of the ice; the initial and final states have the same entropies. According to the law of entropy increase, this implies

$$Z_1 \overset{A}{\sim} Z_2 \tag{3.19}$$

The initial and the final states are adiabatically equivalent. We have thus shown that due to $Z_1 \prec Z_2$ it is in principle possible to adiabatically desalinate salt water. Conversely, due to $Z_1 \succ Z_2$, it is also possible to adiabatically transform water into ice if the considered thermodynamic system comprises an additional amount of salt dissolving in water. It is unlikely that we would have been able to derive such a far-reaching conclusion on the basis of our physical intuition alone.

It is a remarkable consequence of the entropy principle that the knowledge of the entropies of salt water and ice is sufficient to analyze the adiabatic accessibility of all final states of the compound system. It should be emphasized that we have not made any assumptions on how the salt water and ice interact with each other. Entropy connects different systems due to the possibility of energy transfer. A similar situation holds for the value defined in Sect. 1.2.

The property of extensivity is closely related to additivity. For rational values of the scaling parameter t the conditions of additivity and extensivity are mathematically equivalent. Only for irrational values of t are they not completely identical. However, this is a mathematical subtlety which does not affect our considerations.

Before we address the third property of entropy – its concavity – we should briefly comment on the uniqueness of entropy. Can we replace $S(X)$ by another monotonic function, for instance $S^2(X)$ or $\exp[S(X)]$? Wouldn't these functions do the job of describing adiabatic accessibility equally well? This is not the case. Indeed, the latter functions are monotonic but not additive and not extensive. We can demonstrate this by considering a fictitious substance whose entropy is given by the hypothetic equation $S(U, V) = a \times (UV)^{1/2}$ with some constant a. The so-defined entropy function is apparently extensive, because it obeys $S(tU, tV) = tS(U, V)$. But if we introduce a new entropy $S'(U, V) = S^2(U, V)$, then extensivity breaks down due to $S'(tU, tV) = t^2a^2UV = t^2S'(U, V) \neq tS'(U, V)$. This shows that entropy is unique up to a constant, as already mentioned in Sect. 3.1, but it cannot be rescaled arbitrarily.

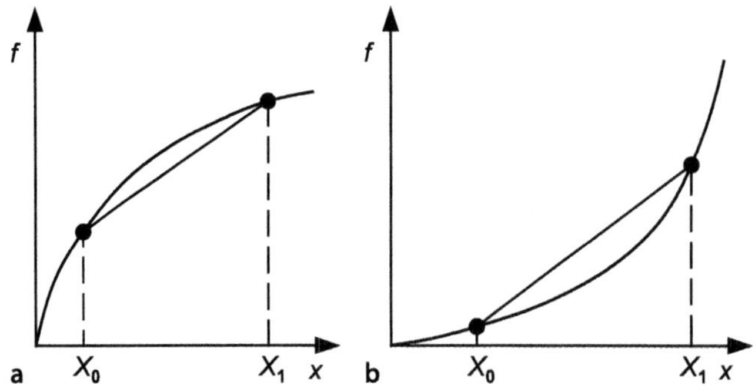

Fig. 3.8.
Concave and convex functions. Example of a concave (a) and convex (b) function of a single variable. The entropy of a simple system is a concave function of the internal energy as explained in Sect. 4.6 and shown in (b). If a function $f(x)$ is twice differentiable, the conditions for concavity and convexity are $\mathrm{d}^2f/\mathrm{d}x^2 < 0$ and $\mathrm{d}^2f/\mathrm{d}x^2 > 0$, respectively

C - Concavity

The properties of entropy discussed so far – monotonicity, additivity, and extensivity – are independent of the geometric features of the state space in which X, Y, and Z are defined. The explanation of concavity, which plays an important role for the definition of temperature, relies on the introduction of coordinates. Let us therefore consider a simple system with one work coordinate. As was already shown in Sect. 2.5, such a system is characterized by the two-dimensional state vector $X = (U, V)$.

For a function $f(x)$ that depends on one variable only, the properties of concavity and convexity are easy to understand. Roughly speaking, $f(x)$ is concave if the line connecting two arbitrary points x_0 and x_1 is located below $f(x)$ as is shown in Fig. 3.8a. In the opposite case which is shown in Fig. 3.8b, the function $f(x)$ is convex. The entropy principle states that entropy is a concave function of the coordinates.

In order to understand the physical content of the abstract expression $S(tX + (1-t)Y) \geq tS(X) + (1-t)S(Y)$ appearing in the entropy principle, let us assume that $X = (U_x, V)$ and $Y = (U_y, V)$ are two states of one kilogram of liquid water with the same volume but different internal energies. We shall further stipulate that the entropies of the states X and Y are S_x and S_y, respectively. Moreover, we shall restrict our attention to the case $t = 1/2$. The statement that entropy is concave is equivalent to saying that the combination of 1/2 kg of water in state X and 1/2 kg of water in state Y to a new system consisting of one kilogram of water in the state $Z = X/2 + Y/2 = (U_x/2 + U_y/2, V)$ is an adiabatic process. Consequently, the entropy of the new system $S(X/2 + Y/2)$ must be higher than the sum of the entropies in the initial state. For arbitrary values of the parameter t within the interval $0 \leq t \leq 1$, the entropy as a function of U must have the same structure as $f(x)$ in Fig. 3.8a.

Chapter 4 General Conclusions

Chapter 4

General Conclusions

As a result of our considerations in the previous chapter, we have learned that each thermodynamic system is characterized by an entropy function $S(X)$ which is monotonic, additive and concave. For a simple system with one work coordinate V, which represents the elementary building block of thermodynamics, the entropy $S(U, V)$ is thus a real function of two variables. In the present chapter, we shall derive general conclusions that are independent of the specific form of this function. In Chap. 5 we will apply the conclusions to specific thermodynamic systems.

4.1 Irreversible and Reversible Processes

Irreversibility and *reversibility* are often considered to be basic concepts of thermodynamics on which the definition of entropy is built. This view is not correct. The basic elements of thermodynamics, also called *primitive concepts*, are equilibrium states and adiabatic accessibility. They enable us to define entropy which in turn provides us with the possibility to define irreversible and reversible processes.

Intuitively, we say that a process is irreversible if it is impossible to return the system *and its surroundings* from the final state back to the initial state. In the opposite case, we call the process reversible. How can we translate this fuzzy description into the accurate language of mathematics?

Let us try the following definition: A process transforming a thermodynamic system from state X with entropy $S(X)$ to state Y with entropy $S(Y)$ is called irreversible if $S(X) < S(Y)$ and reversible if $S(X) = S(Y)$. Such a definition does not make sense for the following two reasons. First, it leaves an open question, namely how we can distinguish between irreversible and reversible processes in the case of decreasing entropy. Second, the properties of the surroundings of the system, which we regard as crucial, do not enter the definition.

The following simple example shows us how serious the second drawback is. The entropy of an ice cube increases if it is transformed into a liquid state. But it would be incorrect to interpret this entropy increase as a signature of an irreversible process. If the ice cube melts as a result of energy transfer in a microwave oven as shown in Fig. 2.1a, we would indeed have an irreversible process. This is reflected in the relation $X \prec\prec Y$, which we have found in Sect. 2.3 for an ice cube melting in a microwave oven. Remember that $X \prec\prec Y$ means that it is impossible to take the system back from Y to X without changing something in the surroundings of the system. By contrast, if the ice cube melts as a result of energy exchange with hot water that is accomplished by some idealized engine as in Fig. 2.1c (see also Fig. 3.3a, case c), then the melting process is reversible! If we include the surroundings of the system into our consideration, namely the hot water in the initial state X_s and final state Y_s, we arrive at the conclusion $(X, X_s) \overset{A}{\sim} (Y, Y_s)$. The adiabatic equivalence expresses the fact that the compound system consisting of the ice cube and its surroundings – the hot water – can be transformed back and forth between its initial and final states and that this process is reversible. These considerations show us that the entropy increase in a melting ice cube does not automatically imply that the melting process is irreversible. A definition of irreversibility on the sole basis of the state of a system is therefore impossible.

A consistent mathematical definition of irreversibility must include both the system and its surroundings. Furthermore, it has to rely on the concept of adiabatic accessibility. The correct definition proceeds as follows. If the system and its surroundings in the initial state (X, X_s) is adiabatically equivalent to the final state (Y, Y_s), the process can apparently be reversed by raising or lowering a weight. Then we have a reversible process. Taking into account the entropy principle, its mathematical property is $S(X, X_s) = S(Y, Y_s)$. Conversely, an irreversible process is obviously characterized by $(X, X_u) \prec\prec (Y, Y_u)$. By virtue of the entropy principle, this condition is equivalent to $S(X, X_s) < S(Y, Y_s)$. The case of decreasing entropy $S(X, X_s) > S(Y, Y_s)$ cannot occur. Indeed, the compound system consisting of the considered thermodynamic system and its surroundings always undergoes an adiabatic process. According to the law of entropy increase, the entropy cannot decrease for such a process. This leads us to the important conclusion that irreversibility and reversibility can only be defined for adiabatic processes. We can then denote the states (X, X_s) and (Y, Y_s) of the compound system simply as X and Y and formulate our definition as follows:

Irreversible process. An adiabatic process that takes a system from state X to state Y is called irreversible if the entropy of the system increases, i.e., $S(X) < S(Y)$.

Reversible process. An adiabatic process that takes a system from state X to state Y is called reversible if the entropy of the system is unchanged, i.e., $S(X) = S(Y)$.

With these two definitions, we have put the terms irreversibility and reversibility on a solid mathematical basis. Let us now verify that these definitions indeed make sense. In Sect. 1.1, we discussed the fall of a stone into a well as an intuitive example of an irreversible process. Consider the water as a thermodynamic system and let us investigate the character of its transition from the initial to the final state. The potential energy of the stone will have decreased after its landing in the water. The internal energy of the water will have increased by the same amount which leads to a slight increase of the water temperature. It is obvious that the water has undergone an adiabatic process, since the only change outside the system is that a body – the stone – has changed its position in the gravity field of the Earth (cf. the definition of adiabatic accessibility and adiabatic processes in Sect. 2.3). According to the entropy principle the entropy of the water is a monotonically increasing function of the internal energy. Since the latter has increased, the entropy must have grown as well. Thus we conclude that the fall of a stone into a well is an irreversible process. The law of entropy increase states that the entropy cannot decrease in adiabatic processes. This law has now a clear correspondence with our everyday experience that water in a well cannot spontaneously cool and throw a stone into the sky.

A second example for an irreversible process is the equilibration of temperatures shown in Fig. 2.1b. Here we have a compound system whose initial state consists of a hot and a cold body and whose final state comprises two bodies with identical temperatures. The equilibration of temperature is an adiabatic process because the surroundings of the system do not change. The hot body reduces its internal energy by the amount ΔU, whereas the internal energy of the cold body rises by ΔU. By virtue of the concavity of entropy, which we shall further elaborate in Sect. 4.3, the entropy increase of the cold body exceeds the entropy decrease of the hot body so that the entropy of the compound system rises. The impossibility of adiabatic processes with decreasing entropy, which is a consequence of the law of entropy increase, manifests itself in our everyday experience that two bodies with identical temperatures cannot spontaneously develop different temperatures.

As a third example, we consider the compression of a gas which is sketched in Fig. 2.2a and b. Let us assume that the lifting device is frictionless and the movement of the piston does not involve any friction, either. We have thus an adiabatic process. If we could calculate the entropies of the initial and final states (which we will do in Sect. 4.7), we would arrive at the conclusion that in general $S(X) < S(Y)$ and the process is irreversible. The reason for the irreversibility is that the piston induces a flow inside the gas whose kinetic energy is gradually dissipated by viscous friction. After reaching the equilibrium state Y, the gas will have a higher internal energy than in the case of a very slow compression. If we could accomplish the compression infinitely slowly, there would be no internal friction in the gas and we would obtain $S(X) = S(Y)$, which indicates a reversible process. This example shows us that reversible processes are idealiszd limiting cases of real (irreversible) processes.

Thanks to the introduction of entropy, we have succeeded in defining a sharp boundary between reversible and irreversible processes. The decision to which class a given process belongs has thus transformed from a matter of speculation into a simple mathematical calculation. The existence of entropy also enables us to tell thermodynamic irreversibility from other forms of irreversibility and subjective perceptions. The destruction of a new car in a car crusher may appear to us as an incarnation of an irreversible process. Thermodynamically, however, it is neither a reversible nor an irreversible process because a car is not scalable and, strictly speaking, does not represent a thermodynamic system in the sense of our definition from Sect. 2.1. The breakdown of a friendship, the loss of property in gambling, or the ageing of a person are certainly irreversible. However these are not irreversible processes in the sense of thermodynamics because "friendship," "property" and "person" do not represent thermodynamic systems in the sense of our definition for which entropy could be defined.

Finally a word of caution is in order concerning the popular claim that the entropy in the office of a disorganized professor increases continuously. If, for simplicity, we consider the professor's office as a compound system consisting of 500 kg of paper and 50 kg of air, we can quickly convince ourselves that its entropy is entirely independent of the arrangement of the subsystems, such as the single sheets of paper. The increasing disorder in the office is therefore not an irreversible process, at least not in the sense of thermodynamics.

4.2 Thermal Equilibrium and Temperature

When formulating the entropy principle, we did not make use of the terms temperature or heat. It was not even necessary to use the expressions *cold* or *hot* to define entropy. We have avoided these terms because they are not only unnecessary for the definition of entropy but also because they are even detrimental to a good understanding of the central role of entropy: "Temperature … is a corollary of entropy; it is epilogue rather than prologue" (Lieb and Yngvason 1999). This will be shown in this chapter as well as in the following chapters.

Consider a compound system consisting of two subsystems. Each subsystem is supposed to be a simple system having a single work coordinate. Let the subsystems be described by the entropies $S_1(X_1)$ and $S_2(X_2)$. Assume that initially each of the subsystems is in equilibrium, as shown in Fig. 4.1a. In order to be specific, let us suppose that the initial state $X_1 = (U_1, V_1)$ describes 2 kg of ice, whereas $X_2 = (U_2, V_2)$ corresponds to 1 kg of liquid lead. We now define a new thermodynamic system, the *thermal join*. We can think of this thermal join as a new system that is obtained by connecting the two systems with a copper wire shown in Fig. 4.1. The thermal join is a simple system because it has only one energy coordinate, namely $U = U_1 + U_2$. The individual internal energies U_1 and U_2 can no longer be independently varied. The mathematical definition of a thermal join is of course independent of such pictures as "ice," "lead" and "copper wire" and reads as follows.

> **Thermal join.** The thermal join of two simple systems in the states $X_1 = (U_1, V_1)$ and $X_2 = (U_2, V_2)$ is a simple system in the state $Y = (U, V_1, V_2)$ whose energy is the sum of the energies of the subsystems, i.e., $U = U_1 + U_2$.

The unification of two systems to a thermal join will in general lead to changes in the subsystems. According to our definition of adiabatic processes given in Sect. 2.3, the processes of the formation of a compound system are adiabatic. The process under

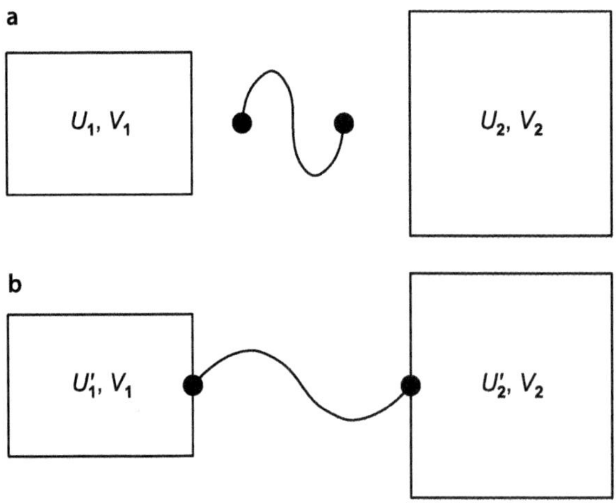

Fig. 4.1. *Thermal join.* Two simple systems (a) are connected using a copper wire and form a thermal join (b). Once the thermal join is created, a relaxation process sets in. During this process, the subsystems exchange energy with each other whereby the total internal energy remains constant and hence $U_1 + U_2 = U_1' + U_2'$. During the relaxation process, the system is not in thermodynamic equilibrium. The relaxation process ends when an equilibrium state is reached. Notice that the thermal join is a simple system, since it possesses only one energy coordinate $U = U_1' + U_2'$

consideration has the particular property that no work is performed. The copper wire leads to a redistribution of the internal energies of the subsystems, but the total energy U remains constant. This process is in general irreversible, as we will see later. Without any detailed computation it is intuitively clear that soon after the installation of the copper wire, the ice will melt and the lead will solidify. A quantitative analysis of the entropy changes in the subsystem would lead to the conclusion that the total entropy of the system increases. Hence the formation of a thermal join is in general connected with entropy production. However, there is one very important exception to this rule! If the initial states of the subsystems are such that after the installation of the copper wire no changes take place, then obviously no entropy will be produced. This case is of crucial importance because it leads us to the concept of thermal equilibrium.

Let us consider the family of all states of the compound system in which the sum of the internal energies of the two subsystems is equal to a given value U. There are obviously infinitely many such states. Indeed, for each possible U_1 there is a U_2, namely $U_2 = U - U_1$ such that the sum of the energies is equal to U. The initial state consisting of ice and liquid lead is just one member of this family. A state consisting of hot water and solid lead represents another example. Crucial for the introduction of the concept of thermal equilibrium is the following question: Which initial state of the compound system does not lead to a redistribution of energy when a thermal join is formed?

According to the entropy principle, no adiabatic process can take place when we form a thermal join unless the total entropy $S = S_1(U_1, V_1) + S_2(U_2, V_2)$ can increase on account of a redistribution between U_1 and U_2. If, however, the total entropy is already at its maximum possible value for a certain value of U_1 (and the complementary value $U_2 = U - U_1$), then the entropy principle leaves no freedom for a further entropy increase. In such a case, the entropy remains constant after the formation of the thermal join. Initial state and final state are then adiabatically equivalent. The unification of the subsystems does not entail any change. The situation can be compared to the behavior of a mechanical system consisting of a ball which is placed into a trough. If it is put into the lowest point of the trough, its potential energy is minimum and the ball will remain at rest. Similar to the definition of mechanical equilibrium of such a ball, we can say that two systems are in thermal equilibrium with each other if their states do not change after formation of a thermal join. This can be mathematically expressed as follows.

> **Thermal equilibrium.** Two simple systems in the states $X_1 = (U_1, V_1)$ and $X_2 = (U_2, V_2)$ are said to be in thermal equilibrium, written $X_1 \overset{T}{\sim} X_2$, if they are adiabatically equivalent to the thermal join $Y = (U_1 + U_2, V_1, V_2)$, i.e. $(X_1, X_2) \overset{A}{\sim} Y$.

What are the particular properties of the thermal equilibrium of two systems? If two systems are in thermal equilibrium, they cannot spontaneously leave this state because the entropy is maximum for the given value of U. In order to determine this maximum, let us compute the change of entropy if the system is slightly displaced from the equilibrium. Let us accomplish this displacement by redistributing the internal energy between the subsystems while keeping their work coordinates constant. The total entropy of the system equals $S = S_1(U_1, V_1) + S_2(U_2, V_2)$. If we change the internal energies of the subsystems by the infinitesimal amounts dU_1 and dU_2, the entropy of the whole system will change by

$$dS = \left(\frac{\partial S_1}{\partial U_1}\right)_{V_1} dU_1 + \left(\frac{\partial S_2}{\partial U_2}\right)_{V_2} dU_2 \tag{4.1}$$

The indices at the brackets indicate that the derivatives have to be computed at constant volume. Since $U = U_1 + U_2$ is constant, we cannot vary the internal energies of the subsystems independently. Rather, they have to satisfy the condition of energy conservation $dU_1 = -dU_2$. Then Eq. 4.1 becomes

$$dS = \left\{\left(\frac{\partial S_1}{\partial U_1}\right)_{V_1} - \left(\frac{\partial S_2}{\partial U_2}\right)_{V_2}\right\} dU_1 \tag{4.2}$$

In order for S to have a maximum, it is necessary that $dS = 0$ for arbitrary values of dU_1. This is impossible unless the term inside the bracket vanishes. The condition for this reads

$$\left(\frac{\partial S_1}{\partial U_1}\right)_{V_1} = \left(\frac{\partial S_2}{\partial U_2}\right)_{V_2} \tag{4.3}$$

With this equation, we have obtained a quantitative criterion for the thermal equilibrium of two systems: Two systems are in thermal equilibrium if their values of $(\partial S/\partial U)_V$ coincide.

The existence of a scalar quantity that quantifies the thermal equilibrium of two systems is a particularly important consequence of the entropy principle. This importance suggests providing this quantity with a special name. We shall refer to its inverse as *temperature* T and define

$$T = \left(\frac{\partial S}{\partial U}\right)_V^{-1} \tag{4.4}$$

This expression looks quite impenetrable and seems to have no relation to our intuitive understanding of temperature. When we will be analyzing the coefficient of performance of heat engines in Sect. 4.5, we will see that this quantity is indeed the temperature we know. The temperature is an intensive quantity, since it is the derivative of an extensive quantity with respect to another extensive quantity, and the ratio of two extensive quantities is an intensive quantity. If two systems in the states X_1 and X_2 are in thermal equilibrium with each other, which we write as $X_1 \overset{T}{\sim} X_2$, then the systems which are by a factor λ_1 and λ_2 larger or smaller, and which are in the states $\lambda_1 X_1$ and $\lambda_2 X_2$, respectively, are also in thermal equilibrium. Hence, the equilibrium condition $X_1 \overset{T}{\sim} X_2$ can be written in the form $T_1(X_1) = T_2(X_2)$.

The temperature defined by Eq. 4.4 has the dimension energy/entropy and thereby the unit Joule/Clausius. Historically, temperature was introduced before entropy. Therefore, the temperature unit *Kelvin* represents the primary SI-unit of thermodynamics whose value at the triple point of water is defined as $T = 273.16$ K. As a consequence, the entropy unit Clausius is defined as 1 Cl = 4.1868 J K^{-1} in the SI-system. For an extensive discussion of the units of temperature and entropy, we refer to textbooks on thermodynamics such as Moran and Shapiro (1995). We can summarize our conclusions regarding temperature as follows.

Temperature. For each simple system with entropy $S(U, V)$, there is an intensive quantity $T = (\partial S / \partial U)_V^{-1}$ called temperature. Two simple systems are in thermal equilibrium if their temperatures are the same.

The equivalence relation $\overset{T}{\sim}$ is transitive because it can be reduced to a comparison of temperatures. If $X \overset{T}{\sim} Y$ and $X \overset{T}{\sim} Z$ hold, then $Y \overset{T}{\sim} Z$ holds as well. Moreover, temperature is a monotonically increasing function of internal energy. The proof of this important theorem is beyond the scope of our book and is not discussed here in detail.

The existence of temperature is sometimes called the zeroth law of thermodynamics. This evokes the impression that it is a fundamental postulate of thermodynamics. As we have seen here, the existence of temperature is a consequence of the entropy principle and not vice versa.

4.3 Heat and Heat Flux

A – Beware of Heat!

"No one has ever seen heat, nor will it ever be seen, smelled or touched." This sentence from Lieb and Yngvason's (1999) original paper contains an important message. "Heat" is certainly important for someone who is about to enter the field of thermodynamics. In a similar way, "horseshoe magnet,", "condenser," "resistor," and "coil" are extremely useful for a beginner entering electrodynamics. However, the accurate mathematical formulation of electrodynamics in the form of Maxwell's equations does not contain any reference to horseshoe magnets. It is based on the fundamental quantities "electric field," "magnetic field," "charge density," and "electric current density," which are connected by partial differential equations. Just as it is not necessary in theoretical electrodynamics to give an exact mathematical definition of a horseshoe magnet, it is difficult and not necessary do define "heat" in an unambiguous manner. The fundamental quantities "energy" and "entropy" as well as the derived quantity "temperature" together with the entropy principle represent the complete "toolkit" that suffices to solve all thermodynamic problems. Heat has played an important role in the history of thermodynamics, but in the accurate mathematical analysis of practical problems of thermodynamics, the use of this term should be avoided. Nevertheless, the word "heat" plays an important role in our everyday life and in the scientific literature. We shall therefore attempt to relate it to the thermodynamic quantities defined here.

Let us first show that it is impossible to unambiguously define "heat." To accomplish this task, we consider a cylinder with variable volume filled with a gas. We assume that the gas undegos a transition from an initial state $X = (U, V)$ to a final state $Y = (U + \Delta U, V)$. We further assume that this transition is accomplished by periodically changing the volume of the gas through fast movements of the piston. At the end of the process, the piston is assumed to return to its initial state and we wait until all flow of the gas has decayed. Since we have accomplished the process in a purely mechanical way, namely by changing the work coordinate, we can say that the increase of the internal energy by the amount $\Delta U = W$ is due to the work W. But we could have argued as well that the work has been converted to heat due to the viscous friction in the gas. In this case, we would have to interpret the increase of the internal energy as a transfer of heat to the system in the amount of $\Delta U = Q$. Obviously we are in a situation where we do not know whether ΔU should be interpreted as heat or work. This example shows that in the general case, when the thermodynamic system is not in equilibrium during the transition from the initial to the final state, it is impossible to uniquely define "heat" by looking at the given system alone.

B – Definition of "Heat" for Arbitrary Processes

If we wish to formulate an unambiguous definition of "heat" for a general thermodynamic process, we need to include the surroundings of the system into our considerations. Remember that for an adiabatic process, the initial and final states of the surroundings differ only in that a weight with mass m may have changed its vertical position in the gravitational field of the Earth by the amount z. Then the work $W = mgz$ (where g is the acceleration of gravity) leads to a change of the internal energy of the system according to

$$U_2 - U_1 = W \tag{4.5}$$

Here U_1 and U_2 are the internal energies of the system before and after the process, respectively. Hence, it is appropriate to denote the change in internal energy in an adiabatic process as work, even if it associated with viscous friction as in the previous example.

After having clarified the character of a process which involves only work, we have to consider processes which involve the exchange only of "heat." To accomplish this task, we have to recall the term *reservoir*, which has been introduced in Sect. 2.5. A reservoir is a simple system whose work coordinate does not change, such as a large block of copper or a vessel filled with compressed gas. If the energy of a reservoir increases by Q, we can say that the reservoir has acquired the "heat" Q. In the opposite case, we say that the reservoir has released the "heat" Q. With this step done, we can define "heat" for an arbitrary processes of a thermodynamic system as follows.

> **"Heat."** If a thermodynamic system undergoes a transition from a state with internal energy U_1 to a state with internal energy U_2 whose only effect is that the internal energy of a reservoir interacting with the system has *decreased* by Q, then the quantity $Q = U_2 - U_1$ is called "heat" transferred from the reservoir to the system. If $Q < 0$, it is said that "heat" has been transferred from the system to the reservoir.

Observe that this definition is not restricted to simple systems. In the case of a compound system, U_1 and U_2 have to be interpreted as the sums of the internal energies of the subsystems before and after the process, respectively.

In the case of an arbitrary process, the initial and final states will in general differ by the position of a weight *and* by the internal energy of a reservoir. Then the energy balance takes the form

$$U_1 - U_2 = W + Q \tag{4.6}$$

This formula is sometimes referred to as the first law of thermodynamics. However, it should be emphasized that the essence of the first law is neither in the symbols $W + Q$ nor in the introduction of "heat." The content of the first law is the observation that the work during an adiabatic process of a thermodynamic system does not depend on the way in which the process is actually carried out (Remember that an adiabatic process is a process where the only difference in the surroundings of a system between the initial and the final state is that a weight has changed its position). This fact has already been mentioned in Sect. 2.5, where we have defined energy coordinates and work coordinates.

C – Definition of "Heat" for Quasistatic Processes

We now turn to the analysis of an important particular case in which work and "heat" can be defined without the need to consider the surroundings of the system. This particular case concerns *quasistatic processes* of simple systems. A quasistatic process is a series of changes of the state of a system during which the system is always in thermodynamic equilibrium. For instance, if we were to compress 1 m^3 of air by 1 mm^3 per day, this would represent a quasistatic process to a very good approximation (However, it would take more than one million years to reduce the volume of the air by the factor of two in this way). Whereas for a general process only the initial state $X_1 = (U_1, V_1)$ and the final state $X_2 = (U_2, V_2)$ are well-defined points in the state space, the states of a quasistatic process form a curve that connects the initial and final states. For a quasistatic process, we can thus consider S, U and V to be continuous variables.

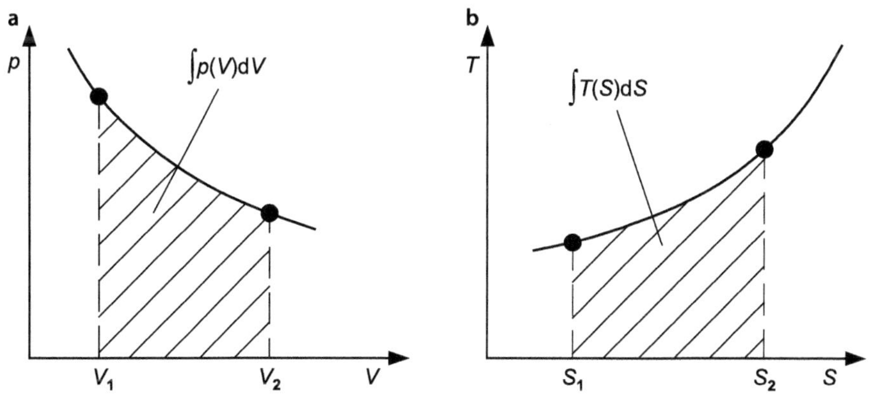

Fig. 4.2. *Work and heat for quasistatic processes.* Geometric interpretation of the work (a) and heat transfer (b) for a quasistatic process as the area in the p-V- and T-S-plane, respectively

In order to uniquely describe a quasistatic process of a simple system, it is necessary to specify the state of the system as a function of a continuous parameter τ with $0 \leq \tau \leq 1$ in the form $X(\tau) = (U(\tau), V(\tau))$ where $X(0) = X_1$ and $X(1) = X_2$. Instead of the parameter τ we can also select a quantity, say V, as an independent variable and specify the remaining two quantities in the form $U(V)$ and $S(V)$.

Let us first define work by considering a quasistatic process with constant entropy. Such a process is called a quasistatic isentropic process. For an infinitesimal variation of V, the internal energy changes by

$$\mathrm{d}U = \left(\frac{\partial U}{\partial V}\right)_S \mathrm{d}V \tag{4.7}$$

The quantity

$$p = -\left(\frac{\partial U}{\partial V}\right)_S \tag{4.8}$$

is referred to as the *generalized pressure*. Integrating Eq. 4.7, we obtain the change of internal energy for a quasistatic process as

$$U_2 - U_1 = -\int_{X_1}^{X_2} p\mathrm{d}V \tag{4.9}$$

This change is equal to the work W. We should be careful in the interpretation of this integral. The pressure depends on the state X and thereby on τ. The symbolic notation $p\,\mathrm{d}V$, which is often used in textbooks of thermodynamics, should therefore be interpreted as an abbreviation for $p(U(\tau), V(\tau))\mathrm{d}V(\tau)$. The integral in Eq. 4.9 is then, strictly speaking, the line integral $W = -\int p(U(\tau), V(\tau))(\mathrm{d}V/\mathrm{d}\tau)\mathrm{d}\tau$. In the case of a compressible substance, the work coordinate V represents the volume in which case the generalized pressure is identical with the usual pressure. Then Eq. 4.9 can be interpreted as compression work as it is done in introductory textbooks of thermodynamics (e.g., Moran and Shapiro 1995). However, Eq. 4.9 holds for a much wider class of systems, including magnetic systems where the magnetization plays the role of the work coordinate.

Next we define “heat” by considering an isochoric process which is a process at constant volume. Here U depends only on S. The change of U during an infinitesimally small variation of entropy is

$$\mathrm{d}U = \left(\frac{\partial U}{\partial S}\right)_V \mathrm{d}S \tag{4.10}$$

The quantity

$$T = \left(\frac{\partial U}{\partial S}\right)_V \tag{4.11}$$

has already been defined as temperature in Sect. 4.2. Integration of Eq. 4.10 yields the change of internal energy as

$$U_2 - U_1 = \int_{X_1}^{X_2} T\mathrm{d}S \tag{4.12}$$

We shall denote this quantity as "heat" Q transferred into the system. The integral should be regarded as an abbreviation for $Q = \int T(U(\tau), V(\tau))(\mathrm{d}S/\mathrm{d}\tau)\mathrm{d}\tau$.

Equations 4.9 and 4.12 admit a clear geometric interpretation which is illustrated in Fig. 4.2. The work performed during a quasistatic process is equal to the area below the curve $p(V)$, whereas the "heat" delivered to the system is equal to the area below $T(S)$.

In order to define W and Q for arbitrary quasistatic processes, we first compute the change of internal energy due to an infinitesimal variation of S and V. The result is

$$\mathrm{d}U = \left(\frac{\partial U}{\partial S}\right)_V \mathrm{d}S + \left(\frac{\partial U}{\partial V}\right)_S \mathrm{d}V \tag{4.13}$$

Using the Definitions 4.8 and 4.11, we can write this relation in the form

$$\mathrm{d}U = T\mathrm{d}S - p\mathrm{d}V \tag{4.14}$$

This equation is called *the first TdS-equation.* Integrating this equation from $\tau = 0$ to $\tau = 1$, we obtain the general result

$$U_2 - U_1 = Q + W \tag{4.15}$$

which, once again, represents the first law of thermodynamics.

D - Heat Transfer

At the end of this section, we would like to briefly discuss the transfer of "heat" between two bodies. Thermodynamically, the term *body* refers to a reservoir, i.e., a thermodynamic system whose work coordinate is held constant and whose remaining single thermodynamic variable is the internal energy U. When two bodies are in thermal contact, energy is always transferred from the hot to the cold body. This seemingly trivial observation can be derived from the entropy principle as will be shown next.

Let us consider two identical bodies in the initial states $X_1 = (U_1, V)$ and $X_2 = (U_2, V)$. Without loss of generality we can assume that $U_1 > U_2$, hence $T_1 > T_2$. If we connect the bodies, say by a copper wire, we create a thermal join. After a sufficiently long time, the thermal join will come to a new equilibrium state. In this state, both bodies have the same temperature T, the total energy will be the same as in the initial state, i.e., $U_1 + U_2 = U_1' + U_2'$ and the total entropy will attain the highest possible value permitted for the given total energy. For a constant work coordinate, the temperature is a unique and monotonically increasing function of the internal energy. Hence, the equality of temperatures of both bodies implies that their internal energies must be equal as well (Notice that this statement is only true for identical bodies!). We denote the value of the internal energies in the final state as $U_1' = U_2' = U$. Since $U_1 + U_2 = 2U$, we have $U = (U_1 + U_2)/2$, i.e., U is the arithmetic mean of the internal energies in the initial state. Thus U is always between U_1 and U_2, from which we conclude $U_1 > U > U_2$. Consequently, the internal energy of the hot body decreases ($U - U_1 < 0$, the hot body loses "heat"), whereas the internal energy of the cold body increases ($U - U_2 > 0$, the cold body gains "heat"). We have thus shown that heat transfer is always from the hot to the cold body. We have demonstrated this statement for identical bodies. The generalization to arbitrary bodies is as follows.

Energy ("heat") flows from hot to cold. Let $X_1 = (U_1, V_1)$ be a state of a simple system and let $X_2 = (U_2, V_2)$ be a state of another simple system. Let T_1 and T_2 be their respective temperatures and assume that $T_1 > T_2$. If $X_1' = (U_1', V_1)$ and $X_2' = (U_2', V_2)$ are two states with the same respective work coordinates as the original states, with the same total energy $U_1 + U_2 = U_1' + U_2'$, and for which the temperatures are equal to a common value T, then *(i)* $U_1' < U_1$ (the hot body loses energy), *(ii)* $U_2' > U_2$ (the cold body gains energy), and *(iii)* $T_1 > T > T_2$ (the final temperature is in between the initial temperatures of the hot and cold body).

The proof of this generalized statement is left to the reader as an exercise. In summary, we have shown in this section, in particular in Sect. 4.3 A, that it is difficult to define "heat" uniquely on the basis of the initial and final states of a thermodynamic system alone. For this reason, it is desirable to avoid the use of this term to the greatest possible extent and to use the primary concepts internal energy, entropy, and temperature instead.

4.4 The Second Law of Thermodynamics

The second law of thermodynamics is known in the following three formulations.

Clausius formulation. No process is possible, the sole result of which is that heat is transferred from a cold body to a hot body.

Kelvin-Planck formulation. No process is possible, the sole result of which is that a body is cooled and work is done.

Carathéodory formulation. In every neighborhood of every equilibrium state there is at least one state which cannot be accessed by an adiabatic process.

The second law of thermodynamics is the most remarkable of all laws of classical physics for several reasons:

- No experiment in a macroscopic system has ever shown a contradiction to the second law;
- "Any reproducible violation of it, however small, would bring the discoverer great riches as well as a trip to Stockholm. The world's energy problems would be solved at one stroke" (Lieb and Yngvason 1999);

- The second law only makes statements about the impossibility, not about the possibility of processes;
- No fundamental law of physics of similar importance is still formulated in words rather than in mathematical language;
- For no other physical law do the textbook definitions show such a large variety and ambiguity.

Historically, the second law of thermodynamics was formulated before entropy was introduced. The Lieb-Yngvason theory enables us to put the logical chain in the right order. We shall indeed show that the foregoing three formulations of the second law can be derived from the entropy principle.

A – The Clausius Formulation of the Second Law

The Clausius formulation of the second law of thermodynamics follows from the fact that the entropy is concave. We illustrate this for the special case of two identical bodies in Fig. 4.3a. Before we give an outline or the proof, we have to replace the sentence given above by an accurate mathematical statement. We shall consider the "body" as a reservoir. Remember that a reservoir is a simple thermodynamic system with entropy $S(U, V)$ whose work coordinate V is kept constant. In order to keep the notation as simple

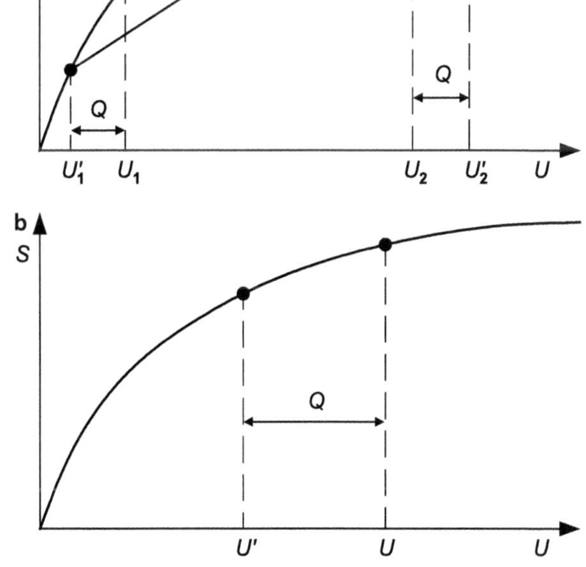

Fig. 4.3. *Second law in the formulation of Clausius, Kelvin and Planck.* (a) Geometric interpretation of the Clausius formulation of the second law of thermodynamics for the particular case of two identical bodies which exchange the energy Q. (b) Geometric interpretation of the Kelvin-Planck formulation of the second law of thermodynamics for a simple system with entropy $S(U)$ when the energy Q is transferred from the system to the surroundings

as possible, we omit the work coordinate and simply write $S(U)$. The term "heat transfer" shall represent a process at the end of which the internal energy of one reservoir decreases whereas the internal energy of another reservoir increases by the same amount.

We assume that in the initial state the reservoirs have the internal energies U_1 and U_2. Without loss of generality we can assume that reservoir 1 has the lower temperature and thus $U_1 < U_2$. If there were a process whose only result would be the transfer of the energy Q ("heat") from the cold to the hot reservoir, then reservoir 1 would have the internal energy $U_1' = U_1 - Q$ whereas reservoir 2 would have the internal energy $U_2' = U_2 + Q$ at the end of the process. In order to demonstrate that such a process is indeed impossible, we need to analyze the total entropy of the compound system consisting of both reservoirs before and after the process. Our goal is to show that this hypothetic process violates the law of entropy increase.

The total entropy of the system in the initial state is

$$S = S(U_1) + S(U_2) \tag{4.16}$$

In the final state, the compound system has the entropy

$$S' = S(U_1') + S(U_2') \tag{4.17}$$

The statement that the heat transfer from the cold to the hot body should be "the sole result" is equivalent to saying that the considered process is adiabatic. The process under consideration represents a particular case because *no* weight has changed its position in the gravitational field (cf. the definition of the adiabatic accessibility in Sect. 2.3). Since the entropy cannot decrease after an adiabatic process, the condition $S \le S'$ must be satisfied. Hence

$$S(U_1) + S(U_2) \le S(U_1') + S(U_2') \tag{4.18}$$

must hold. Figure 4.3a shows that our hypothetic process violates this condition. The quantity $S(U_1) + S(U_2)$ has a simple geometric meaning. It is twice the entropy S_A of the state which is marked with the letter A in Fig. 4.3a. This state is located halfway between U_1 and U_2. Similarly $S(U_1') + S(U_2') = 2S_B$. Since the entropy is a concave function of U, the connecting line between U_1' and U_2' is always located below the line that connects U_1 and U_2 as long as $U_1' < U_1$ and $U_2' > U_2$. But this implies that $S_B < S_A$, i.e., the entropy of our hypothetic process decreases! This contradicts Condition 4.18 which proves Clausius' formulation of the second law of thermodynamics. The proof of the general case of two different bodies is only slightly more complicated and is left as an exercise for the reader. Notice that our proof did not require any explicit form of the function $S(U)$ but only the properties of monotonicity and concavity which are contained in the entropy principle.

B – The Kelvin-Planck Formulation of the Second Law

We now turn to the proof of the Kelvin-Planck formulation of the second law. The Kelvin-Planck version is a consequence of the monotonicity of entropy. As in the previous section, we start by translating the Kelvin-Planck statement into a mathematical language. We shall interpret the words "cooling of a body" as a transition of a reservoir from a state with energy U and entropy $S(U)$ to a state with energy $U' = U - Q$ and entropy $S(U')$. The statement that "work is done" should be understood in such a way that the potential energy of a mass in the gravitational field of the Earth has increased in the sense of the definition of adiabatic accessibility. Let us consider a thermodynamic system whose initial state is characterized by the entropy

$$S = S(U) \tag{4.19}$$

Assume that after the considered process the entropy of the system becomes

$$S' = S(U') \tag{4.20}$$

If "cooling of a body" and "work" are to be the sole results, the process must be adiabatic. Hence, according to the law of entropy increase, the entropy of the whole system must not decrease, i.e., $S' \geq S$. Consequently

$$S(U') \geq S(U) \tag{4.21}$$

must hold. As shown in Fig. 4.3b, however, the entropy is a monotonically *increasing* function of the energy. If the energy decreases by the amount Q, the entropy must decrease as well. However, this would violate the law of entropy increase. Thus we have proved the Kelvin-Planck version of the second law.

C – The Carathéodory Formulation of the Second Law

In 1909, the mathematician Constantin Carathéodory published an article entitled "Investigations into the Foundations of Thermodynamics" (Carathéodory 1909). This

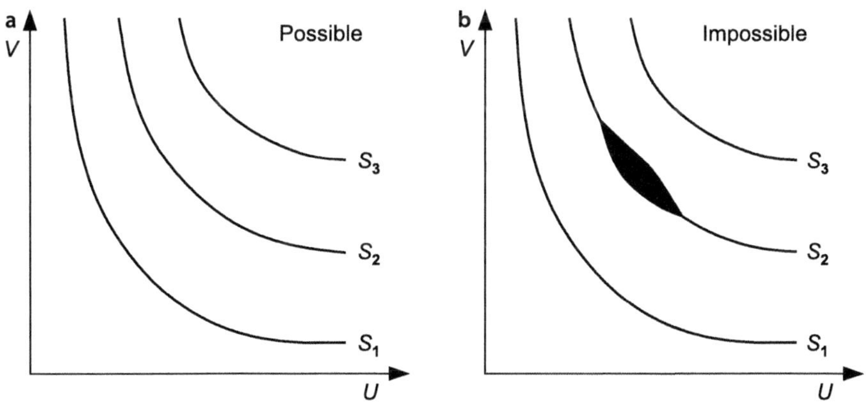

Fig. 4.4. *Second law in the formulation of Carathéodory.* Geometric interpretation of the Carathéodory formulation of the second law of thermodynamics using three sets of points with constant entropies (isentropes) with $S_1 < S_2 < S_3$. The Carathéodory formulation states that the isentropes are always one-dimensional structures as shown in (a). Two-dimensional islands as shown in (b) are impossible

work was to become a milestone in the development of thermodynamics because it showed that the existence of entropy can be derived from what we call today the Carathéodory formulation of the second law of thermodynamics. Lieb and Yngvason (Lieb and Yngvason 1999) have demonstrated in turn that the Carathéodory formulation can be derived from the entropy principle. Since this derivation is quite complex mathematically, we will not repeat it here. We rather confine our attention to providing a geometric interpretation of the Carathéodory formulation.

Figure 4.4 shows the states of a simple system characterized by its internal energy and one work coordinate, namely its volume. The Carathéodory formulation of the second law of thermodynamics states that the points with a given entropy are always one-dimensional manifolds, i.e. lines, such as shown in Fig. 4.4a. Then in each neighborhood of each point there is always at least one point which is not adiabatically accessible. The existence of two-dimensional islands of constant entropy as in Fig. 4.4b is thus excluded by this statement.

4.5
Efficiency of Heat Engines and Refrigeration Cycles

How far can a steam locomotive run with one ton of coal? The general answer to this question was given by the French scientist Sadi Carnot in a fundamental work published in 1824. Carnot did not explicitly use the first law of thermodynamics which expresses energy conservation, neither was the absolute temperature scale known at that time. Nevertheless, he was able to show that the maximum work that can be performed by an engine neither depends on the working fluid (for instance water) nor on the details of the construction but only on the temperatures of the two reservoirs with which the engine interacts. In what follows, we will show that Carnot's central result – the famous formula for the efficiency of a heat engine in terms of temperatures – can be derived from the entropy principle. Moreover, we will show how Carnot's formula can be generalized to refrigeration systems and heat pumps, again using the entropy principle.

A – Efficiency of a Heat Engine

Let us define the term heat engine by making use of Fig. 4.5. Consider two simple systems, each having one work coordinate. Consider further an apparatus, called heat engine, which can interact with these two systems. We denote the system with lower temperature T_C "cold reservoir" and the system with higher temperature T_H "hot reservoir." A heat engine is a device which extracts the energy Q_H from the hot reservoir, performs work W and rejects the energy Q_C to the cold reservoir. The work performed by the engine is symbolically expressed in Figs. 4.5a and b by a weight being lifted. Figures 4.5a and b also contain the values for entropy, energy and temperature at the beginning and at the end of a considered process.

For the process shown in Fig. 4.5, we would like to compute the ratio

$$\eta = \frac{W}{Q_H} \tag{4.22}$$

of the delivered work to the heat taken from the hot reservoir. This ratio is called the thermal efficiency of the process. Since energy is conserved, we have $W = Q_H - Q_C$ and can write the thermal efficiency in the form

$$\eta = 1 - \frac{Q_C}{Q_H} \tag{4.23}$$

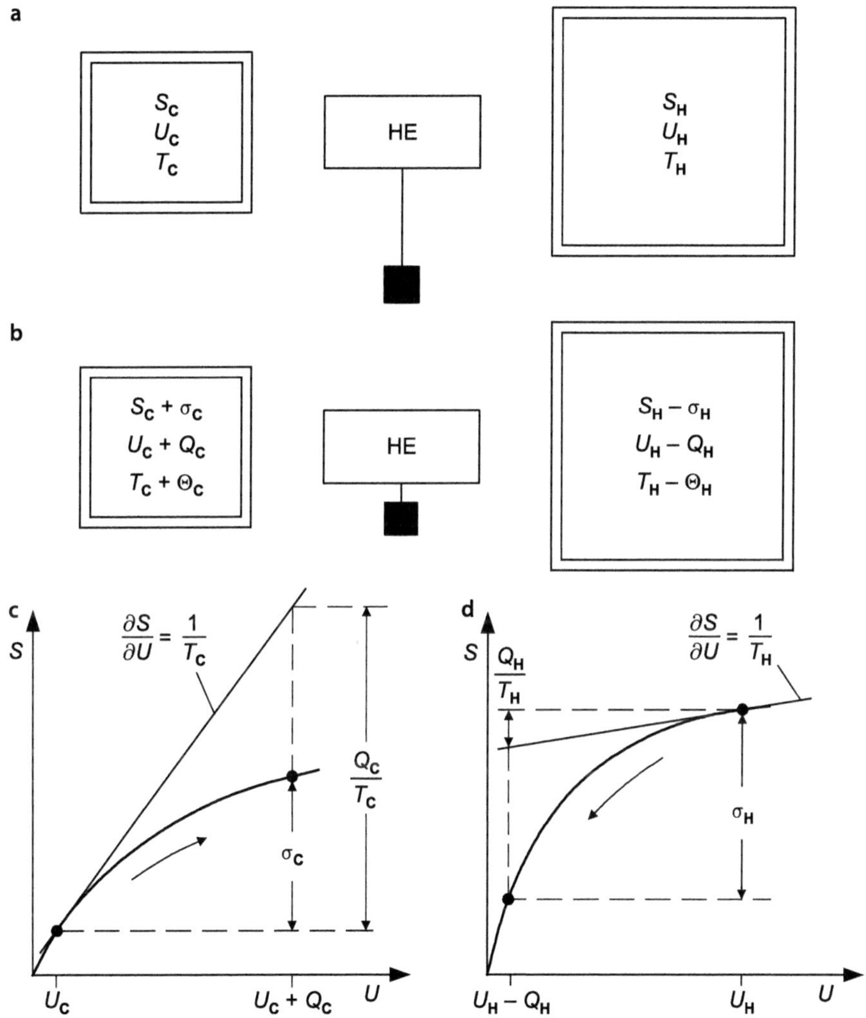

Fig. 4.5. *Efficiency of heat engines.* Entropy, internal energy and temperature of a hot and a cold reservoir (a) before and (b) after their interaction with a heat engine (HE). The indices C and H stand for cold and hot, respectively. (c) Determination of a lower bound for the energy Q_C, (d) determination of an upper bound for the energy Q_H. Notice that the heat capacities of the reservoirs are in general finite and that their temperatures converge after the considered process. However, in the limit of reservoirs with infinite heat capacities, $\Theta_C = \Theta_H = 0$

In the worst case the engine does not deliver any work. Then $Q_C = Q_H$ and $\eta = 0$. In the best case (which is forbidden by the Kelvin-Planck formulation of the second law, see Sect. 4.4 B) the "heat" Q_H would be completely converted into work. Then $Q_C = 0$ and $\eta = 1$. Hence the thermal efficiency has the property $0 \leq \eta \leq 1$. We know that the case $\eta = 1$ cannot be reached in practice and that the maximum efficiency depends on the properties of the reservoirs. We wish to show below how one can estimate the thermal efficiency.

In order to estimate the maximum thermal efficiency for a given pair of reservoirs we have to determine the maximum of W for a given value of Q_H. This task is equivalent to finding a lower bound for the quantity Q_C / Q_H for a given amount of work W. This in turn requires expressing the lower bound of Q_C through T_C and σ_C and the upper bound of Q_H through T_H and σ_H. This will be done next.

To compute the required upper and lower bounds, we make use of the fact that the entropies of both reservoirs are concave functions of their internal energies. From Fig. 4.5c we can infer that the tangent to the entropy function $S(U)$ at $U = U_C$ has the value $(\partial S / \partial U)_V = 1 / T_C$. Since $S(U)$ is concave, its increase $\sigma_C = S(U + Q) - S(Q)$ is always smaller than the increase Q_C / T_C of the tangent at $U = U_C$. This observation can be mathematically expressed as $\sigma_C \leq Q_C / T_C$ and provides us with the lower bound $Q_C \geq \sigma_C T_C$. A similar consideration for the hot reservoir, illustrated in Fig. 4.5d, leads us to the conclusion that $\sigma_H \geq Q_H / T_H$. Hence we obtain the upper bound $Q_H \leq \sigma_H T_H$. If we insert the two bounds into Eq. 4.23, we obtain

$$\eta \leq 1 - \frac{\sigma_C T_C}{\sigma_H T_H} \tag{4.24}$$

We can sharpen this estimate if we remember that the transition from the initial state shown in Fig. 4.5a to the final state shown in Fig. 4.5b is an adiabatic process. According to the law of entropy increase, the entropy cannot decrease for this process. This is equivalent to $\sigma_C - \sigma_H \geq 0$, from which we conclude $\sigma_C \geq \sigma_H$ and finally $\sigma_C / \sigma_H \geq 1$. In the ideal case, corresponding to a reversible process, we have $\sigma_C / \sigma_H = 1$ and our final result becomes

$$\eta \leq 1 - \frac{T_C}{T_H} \tag{4.25}$$

The maximum efficiency $\eta_{Carnot} = 1 - T_C / T_H$ is also called Carnot efficiency. It is reached if the two reservoirs are so large that their temperatures remain virtually unchanged during the process. Equation 4.25 is one of the hallmarks of thermodynamics and can be expressed in words as follows.

> **Carnot-efficiency of a heat engine.** The thermal efficiency of a heat engine which interacts with a hot reservoir with temperature T_H and a cold reservoir with temperature T_C cannot exceed the Carnot-efficiency $\eta_{Carnot} = 1 - T_C / T_H$.

It is remarkable that the maximum efficiency of a heat engine neither depends on its details of construction nor on the material of the reservoirs but only on their temperatures. It is even somewhat misleading to speak of the efficiency of an *engine*, since η is a not a property of an engine but of two reservoirs!

A further interesting property of Eq. 4.25 is worth mentioning. This equation holds no matter whether the considered reservoirs are infinite or finite. In the former case, which represents a special case, the temperatures of the reservoirs remain unchanged during the process. By contrast, in the case of finite reservoirs their temperatures approach each other. Hence, Eq. 4.25 does also hold for the general case of finite reservoirs provided that the temperatures T_C and T_H are correctly interpreted as the temperatures *before* the process.

Our considerations in this chapter are not only useful for the understanding of the efficiency of heat engines. We can use similar lines of reasoning to derive two useful conclusions about the coefficients of performance of refrigeration systems and heat pumps.

B – Coefficient of Performance of a Refrigeration System

In the language of thermodynamics, a refrigeration system is a device which extracts the energy Q_C from a reservoir at low temperature, receives work W from its surroundings and rejects "waste heat" Q_H to another reservoir which is at high temperature. The best known example of a refrigeration system is a household refrigerator which extracts "heat" from its interior that is to be cooled, rejects "waste heat" to the air of the kitchen and uses electricity to drive the compressor. The coefficient of performance of a refrigeration system is defined as the ratio of the "cold" taken from the cold reservoir to the work that has to be performed, namely

$$\beta = \frac{Q_C}{W} \tag{4.26}$$

Taking into account energy conservation, which is expressed as $W = Q_H - Q_C$, we can write β in the form

$$\beta = \left(\frac{Q_H}{Q_C} - 1\right)^{-1} \tag{4.27}$$

If the refrigeration system required no work, we would have $Q_C = Q_H$ and thus $\beta \to \infty$. If the system would not cool at all, i.e., $Q_C = 0$, the coefficient of performance would be zero. Consequently, β of a refrigeration system must be in the range $0 \leq \beta < \infty$. We would like to know the maximum possible coefficient of performance of a refrigeration system for given initial temperatures T_C and T_H of the cold and hot reservoirs, respectively. To accomplish this task, we have to determine the minimum of W for a given value of Q_C. This is equivalent to placing a lower bound on the quantity $(Q_H/Q_C) - 1$ for a given W. This task in turn requires expressing the lower bound for Q_H as well as the upper bound for Q_C through T_H and σ_H as well as T_C and σ_C, respectively. The necessary estimates can be performed with reference to Fig. 4.6

Figures 4.6a and b show the internal energies, entropies and temperatures before and after a hypothetic process which consists of generating a certain amount of "cold." From Fig. 4.6c we can infer that by virtue of the concavity of the entropy the inequality $\sigma_C \geq Q_C/T_C$ holds. This relation immediately yields the desired upper bound $Q_C \leq \sigma_C T_C$. An analogous consideration for the hot reservoir using Fig. 4.6d provides the result

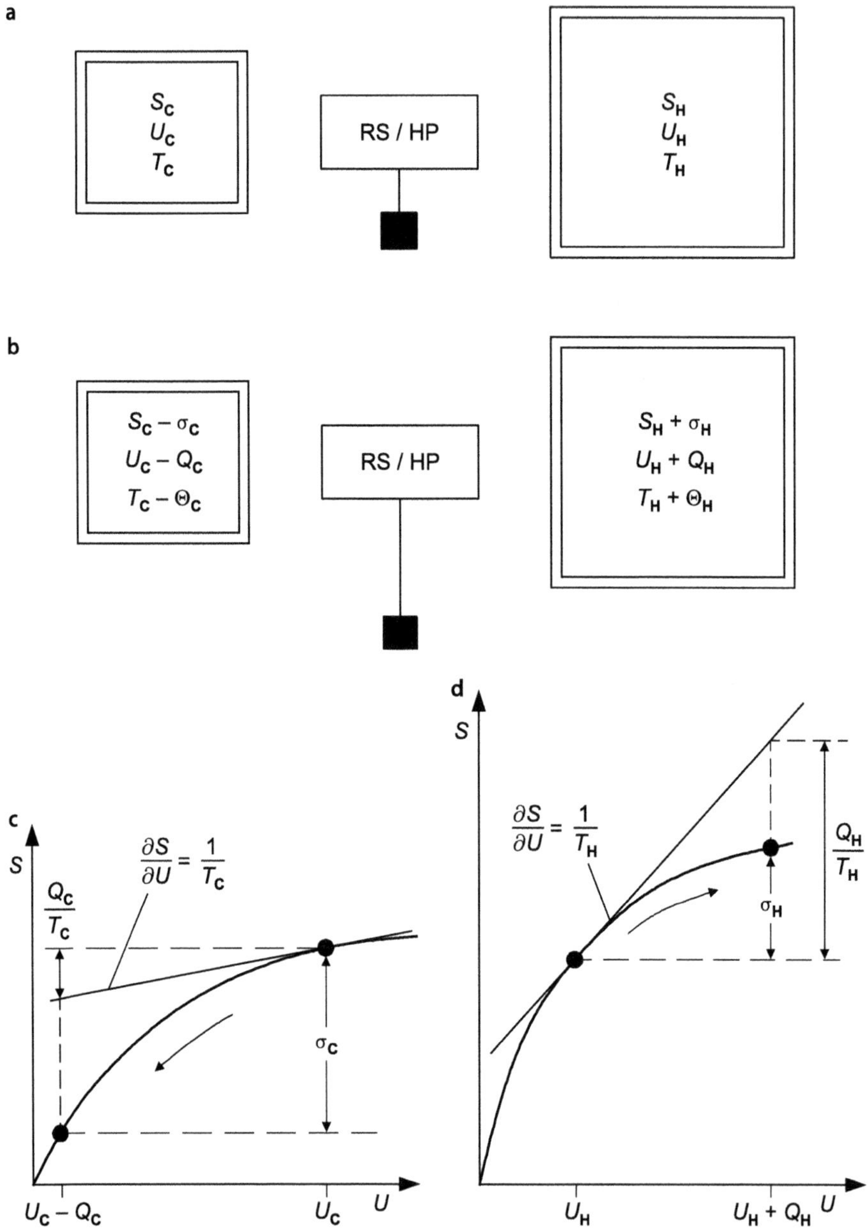

Fig. 4.6. *Coefficient of performance of refrigeration systems and heat pumps.* Entropy, internal energy and temperature of a hot and a cold reservoir (a) before and (b) after their interaction with a refrigeration system (RS) or a heat pump (HP). (c) Determination of an upper bound for Q_C, (d) determination of a lower bound for Q_H. Notice that the heat capacities of the reservoirs are in general finite and that their temperatures diverge after the considered process. However, in the limit of reservoirs with infinite heat capacities, $\Theta_C = \Theta_H = 0$

$\sigma_H \leq Q_H/T_H$ and the lower bound $Q_H \geq \sigma_H T_H$. Inserting the obtained bounds into Eq. 4.27, we are led to the result

$$\beta \leq \left(\frac{\sigma_H T_H}{\sigma_C T_C} - 1\right)^{-1} \tag{4.28}$$

As in the previous section we can sharpen the estimate if we remember that, by virtue of the entropy principle, the entropy production $\sigma_H - \sigma_C$ of a system undergoing an adiabatic process must not be negative. This is mathematically expressed as $\sigma_H \geq \sigma_C$. In the best case, which corresponds to a reversible process, we have no entropy production at all and thus $\sigma_H = \sigma_C$. After a short computation we are led to the final result

$$\beta \leq \frac{T_C}{T_H - T_C} \tag{4.29}$$

The maximum coefficient of performance $\beta_{\text{Carnot}} = T_C/(T_H - T_C)$ is attained for infinitely large reservoirs in which case the temperatures T_C and T_H remain unchanged in the course of the process. Our conclusions can be summarized as follows.

Carnot-efficiency of a refrigeration system. The coefficient of performance of a refrigeration system which interacts with reservoirs with temperatures T_C and T_H cannot exceed the Carnot-efficiency $\beta_{\text{Carnot}} = T_C/(T_H - T_C)$.

The Carnot-efficiency depends neither on the refrigerant nor on the constructive details of the refrigeration system but only on the temperatures of the reservoirs. Equation 4.29 holds both for finite and infinite reservoirs provided that the temperatures appearing in this formula are correctly interpreted as the temperatures at the beginning of the process.

C – Coefficient of Performance of a Heat Pump

As a last application in this chapter, let us analyze the coefficient of performance of a heat pump. A heat pump is a device which, in a given unit of time, extracts the energy Q_C from a cold reservoir, receives work W from its surroundings and supplies the energy Q_H to a hot reservoir. Heat pumps can be used for heating purposes. The application of a heat pump requires mechanical or electrical energy to drive a compressor as well as a reservoir whose temperature is as close as possible to the temperature of the building to be heated. From a thermodynamic point of view, there is no difference between a heat pump and a refrigeration system. They differ in that the utility of a refrigeration system is measured by its refrigeration capacity Q_C whereas the utility of a heat pump is measured by the delivered "heat" Q_H. Due to the similarity of both types of systems, Fig. 4.6 applies both to refrigeration systems and heat pumps. The coefficient of performance of a heat pump is defined as

$$\beta = \frac{Q_H}{W} \tag{4.30}$$

By virtue of energy conservation ($W = Q_H - Q_C$), this equation becomes

$$\beta = \left(1 - \frac{Q_C}{Q_H}\right)^{-1} \tag{4.31}$$

The coefficient of performance has the property $1 \leq \beta < \infty$. In the worst case $\beta = 1$ the heat pump makes no use of the cold reservoir and only converts electricity into "heat." (It is then an electric heater rather than a heat pump.) The best case $\beta \to \infty$, which is forbidden by the Clausius formulation of the second law, corresponds to a heat transfer from the cold to the hot reservoir that does not require any work. If we wish to determine the maximum possible value of β for reservoirs with given initial temperatures, we have to compute the minimum of W for a given value of Q_H. This problem is equivalent to finding a lower bound of $1 - Q_C/Q_H$ for a given W. This in turn requires expressing an upper bound for Q_C and a lower bound for Q_H through T_H and σ_H. This task has already been accomplished for the refrigeration system by invoking Figs. 4.6c and d. There the result was $Q_C \leq \sigma_C T_C$ and $Q_H \geq \sigma_H T_H$. Since the law of entropy increase applies to refrigeration systems and to heat pumps in the same way, the relation $\sigma_H \geq \sigma_C$ must hold. In the idealized case of a reversible process, we have $\sigma_H = \sigma_C$ and we are led to the final result

$$\beta \leq \frac{T_H}{T_H - T_C} \tag{4.32}$$

The maximum efficiency $\beta_{Carnot} = T_H/(T_H - T_C)$, which is also called Carnot-efficiency, is attained if the process is reversible and the reservoirs are infinitely large. Equation 4.32 shows that a heat pump is most efficient when the temperatures of energy supply and energy rejection are as close as possible to each other. Our conclusions can be summarized as follows:

> **Carnot-efficiency of a heat pump.** The coefficient of performance of a heat pump operating between a cold reservoir with temperature T_C and a hot reservoir with temperature T_H cannot exceed the Carnot-efficiency $\beta_{Carnot} = T_H/(T_H - T_C)$.

4.6 Fundamental Thermodynamic Functions

If the concept of entropy were unknown, we would need two separate pieces of information to describe the thermodynamic behavior of a simple system with one work coordinate. These are the thermal equation of state $p(V, T)$ and the caloric equation of state $U(V, T)$. With the help of entropy, the same information can be reduced to a single equation $S(U, V)$. When describing the proof of this statement, we will encounter four functions, namely $U(S, V)$, $H(S, p)$, $F(T, V)$, and $G(T, p)$ which are called *fundamental thermodynamic functions*. This section is devoted to the derivation and interpretation of these functions. Before elaborating further on this topic, we must briefly discuss an important property of the internal energy, namely its convexity.

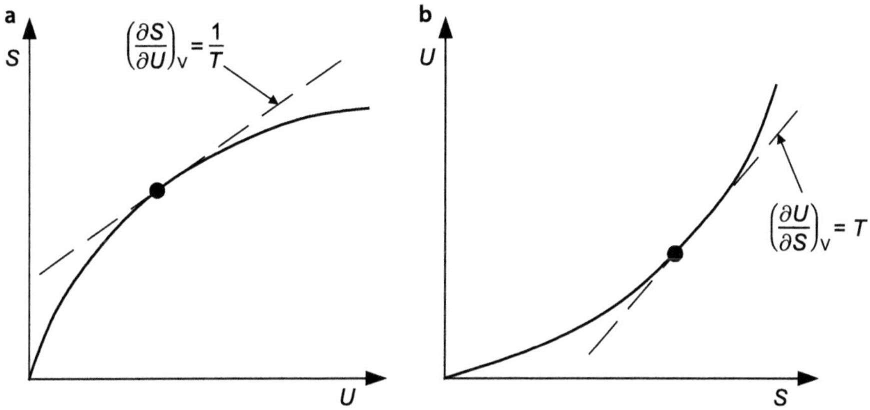

Fig. 4.7. *Convexity of the internal energy.* The entropy $S(U, V)$ of a simple system is a concave function of the internal energy as shown in (a). The internal energy $U(S, V)$ is a convex function as shown in (b)

We start our analysis by considering a simple system with one work coordinate whose entropy is given by $S(U, V)$. Figure 4.7a qualitatively shows how the entropy depends on the internal energy when the work coordinate is held constant. According to the entropy principle, this function is monotonic and concave. As explained in Sect. 4.2, its slope is equal to the inverse of the temperature. On account of the monotonicity and concavity of the entropy, each value of U corresponds to exactly one value of S. Hence we can write the function $S(U, V)$ in the form $U(S, V)$ as well. This is shown in Fig. 4.7b. From this figure, we can infer that the concavity of entropy implies convexity of internal energy as a function of entropy. Indeed, the tangent in Fig. 4.7b, whose slope is equal to the temperature, is always located below $U(S, V)$. We will make use of this characteristic property shortly. First, however, let us define the term fundamental thermodynamic function for a simple system with one work coordinate.

Fundamental thermodynamic function. A function of two thermodynamic variables is called fundamental thermodynamic function if it contains all thermodynamic information about a given simple system with one work coordinate, in particular if the thermal and caloric equations of state can be derived from this function.

A – Internal Energy

We start by demonstrating that $U(S, V)$ is a fundamental thermodynamic function. As a first step let us summarize what we already know about it. The total differential of internal energy is

$$dU = \left(\frac{\partial U}{\partial S}\right)_V dS + \left(\frac{\partial U}{\partial V}\right)_S dV \qquad (4.33)$$

According to the Definitions 4.4 and 4.8 temperature and pressure are given by

$$T = \left(\frac{\partial U}{\partial S}\right)_V, \quad p = -\left(\frac{\partial U}{\partial V}\right)_S \tag{4.34}$$

Hence

$$\mathrm{d}U = T\mathrm{d}S - p\mathrm{d}V \tag{4.35}$$

We have already encountered this equation, called the $T\mathrm{d}S$-equation, in Sect. 4.3 C. If we assume that $U(S, V)$ is twice continuously differentiable, the order of differentiation can be inverted. This is expressed as $\partial^2 U/\partial S \partial V = \partial^2 U/\partial V \partial S$. If we differentiate the first Eq. 4.34 with respect to V and the second Eq. 4.34 with respect to S and make use of this equality we can conclude

$$\left(\frac{\partial T}{\partial V}\right)_S = -\left(\frac{\partial p}{\partial S}\right)_V \tag{4.36}$$

This equation is called Maxwell relation and should not be confused with Maxwell's equations of electrodynamics. In a second step we write the expression $U(S, V)$ in an explicit form as

$$U = f_U(S,V) \tag{4.37}$$

assuming that the function f_U is known for a given system. This notation appears a little bit heavy, but we gain clarity by using it. In a similar explicit notation, temperature and pressure become $T = f_T(S, V)$ and $p = f_p(S, V)$ where f_T and f_p are given by

$$f_T(S,V) = \left(\frac{\partial f_U}{\partial S}\right)_V \tag{4.38}$$

and

$$f_p(S,V) = -\left(\frac{\partial f_U}{\partial V}\right)_S \tag{4.39}$$

In a third step, we symbolically solve Eq. 4.38 for S and obtain

$$S = f_T^{-1}(T,V) \tag{4.40}$$

Here f_T^{-1} is a symbolic notation for the inverse function of f_T at constant V. We can now express the thermal and caloric equations of state in the form

$$p(T,V) = f_p\left[f_T^{-1}(T,V),V\right] \tag{4.41}$$

and

$$U(T,V) = f_U\left[f_T^{-1}(T,V),V\right] \tag{4.42}$$

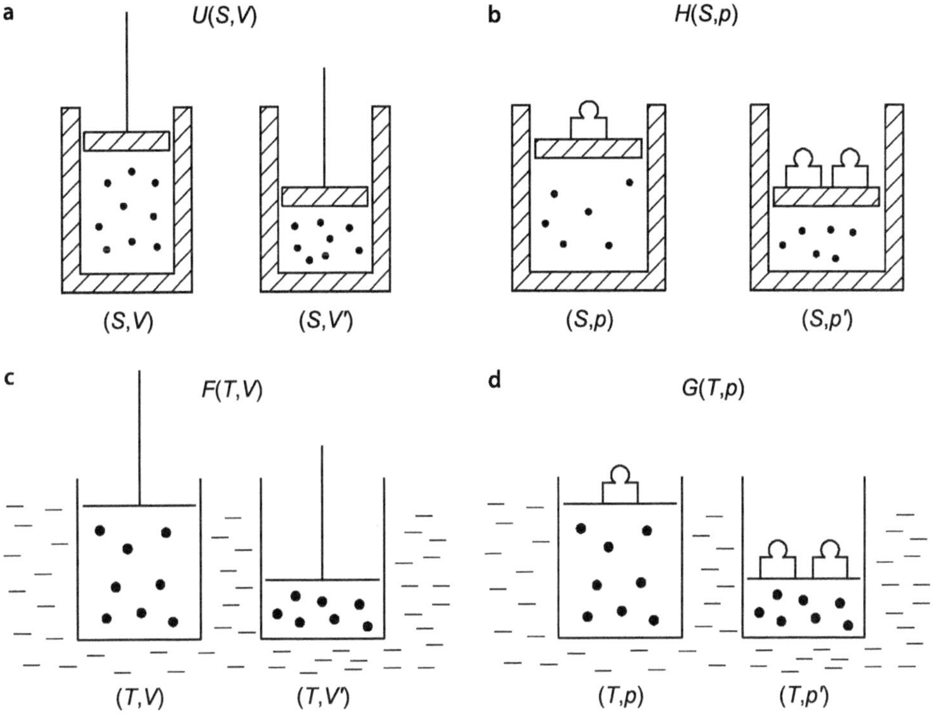

Fig. 4.8. *Fundamental thermodynamic functions and processes.* Overview of important types of thermodynamic processes and the fundamental thermodynamic functions which are best suited for their analysis. (a) Isentropic process with prescribed change of volume and its analysis using internal energy; (b) isentropic process with prescribed change of pressure and its analysis using enthalpy; (c) isothermal process with prescribed change of volume and its analysis using the Helmholtz function; (d) isothermal process with prescribed change of pressure and its analysis using the Gibbs function

We have thus demonstrated that $U(S, V)$ is indeed a fundamental thermodynamic function. Notice that it is not in general correct to say that internal energy is a fundamental thermodynamic function. This statement is incomplete unless further information is provided about the variables on which the function depends. For instance, $U(T, V)$ is not a fundamental thermodynamic function because it is impossible to derive the thermal equation of state from this representation. A correct statement would therefore be that the internal energy as a function of entropy and volume is a fundamental thermodynamic function.

After having characterized $U(S, V)$ as a fundamental thermodynamic function, we wish to illustrate its physical meaning using Fig. 4.8a. The function $U(S, V)$ is particularly useful if one has to compute the pressure and temperature of a system after an isentropic compression or expansion from an initial volume V to a final volume V'. An isentropic compression is a process where the entropies of the initial and final states are equal. We can imagine the particular case of a *quasistatic* isentropic process by picturing an infinitely slow compression of a gas in a thermally well insulated vessel. After the compression, pressure and temperature are given by $p' = -(\partial U/\partial V)_S$ and

$T' = (\partial U / \partial S)_V$. Here the partial derivatives have to be computed for (S, V'). In this way one could for instance compute the pressure and the temperature of the air compressed inside a Diesel engine. Our conclusions about the internal energy can be summarized as follows.

> **Internal energy $U(S,V)$ as a fundamental thermodynamic function.** The internal energy as a function of entropy and volume is a fundamental thermodynamic function which permits one to compute the pressure $p' = -(\partial U / \partial V)_S$ and temperature $T' = (\partial U / \partial S)_V$ of a system after a transition from a state with entropy S and volume V to a state with the same entropy and volume V'.

B – Enthalpy

It is sometimes difficult or even impossible to solve a given practical problem using the internal energy as a fundamental thermodynamic function. For example, when using internal energy it is quite cumbersome to compute the final volume of one cubic meter of air after it has been compressed to a pressure thirty times the ambient pressure when entering the gas turbine of an aircraft. In a highly simplified manner, this problem is represented in Fig. 4.8b. If we were to perform this computation on the basis of the function $U(S, V)$ under the assumption of an isentropic compression, we would have to compute the final volume by determining the value V' for which the derivative $-(\partial U / \partial V)_S$ is equal to a given value of the pressure p'. This would be a tedious procedure. It would be much easier if we had a fundamental thermodynamic function in which the pressure p appears as an independent variable in place of V.

Mathematicians have known for a long time how to replace the independent variable x of a convex function $y = f(x)$ by a new independent variable $\xi = f'(x)$. This task is accomplished using a Legendre transform. This transform does not only lead to a new independent variable but also to a new function. In the present case, this function corresponds to a new fundamental thermodynamic function which is called *enthalpy* $H(S, p)$.

The Legendre transform maps a convex function $y = f(x)$, such as the function shown in Fig. 3.8b, onto another function $\eta = g(\xi)$ which is defined by $\xi = f'$ and $g = f - f' \cdot x$. Simply speaking, after a Legendre transform a function is no longer described by the coordinates (x, y) of its points but rather by its slope ξ and by the point η at which a tangent to the curve at (x, y) intersects the y-axis. For instance, it can be verified by straightforward computation that the Legendre transform of the function $y = x^2$ is $\eta = -\xi^2 / 4$. An important mathematical property is that the Legendre transform can be uniquely inverted.

Returning to thermodynamics, let us perform a Legendre transform of the internal energy with respect to the variable V. This implies that we define a new function, the enthalpy H, which has the form

$$H = U - \left(\frac{\partial U}{\partial V}\right)_S V \qquad (4.43)$$

At first glance this definition looks like a mathematical exercise. After a deeper analysis it turns out that this definition is useful, however. By virtue of our Legendre transform 4.43, we have replaced the volume by the pressure as an independent quantity. We can thus write the enthalpy in the form

$$H = U + pV \tag{4.44}$$

It is not immediately obvious why the enthalpy is a function of S and p only. Doesn't it depend on the volume as well, which appears explicitly in Eqs. 4.43 and 4.44? In order to demonstrate how $H(S,p)$ can be constructed from the knowledge of $U(S, V)$, we return to the explicit notation $U - f_U(S, V)$. We now symbolically invert $p = f_p(S, V)$ and write $V = f_p^{-1}(S,p)$. Finally, we write Eq. 4.44 in the form $H(S,p) = U(S, f_p^{-1}(S,p)) + pf_p^{-1}(S,p)$, which proves our assertion. The differential of the enthalpy is

$$\mathrm{d}H = \left(\frac{\partial H}{\partial S}\right)_p \mathrm{d}S + \left(\frac{\partial H}{\partial p}\right)_S \mathrm{d}p \tag{4.45}$$

At the same time Eqs. 4.35 and 4.44 imply

$$\mathrm{d}H = T\mathrm{d}S + V\mathrm{d}p \tag{4.46}$$

This equation is called the second $T\mathrm{d}S$-equation. By comparing the coefficients in front of $\mathrm{d}S$ and $\mathrm{d}p$ in Eqs. 4.45 and 4.46, we obtain

$$T = \left(\frac{\partial H}{\partial S}\right)_p, \quad V = \left(\frac{\partial H}{\partial p}\right)_S \tag{4.47}$$

For the same reason as in the previous subsection, the order of differentiation can be changed when H is differentiated twice. Hence $\partial^2 H / \partial S \partial p = \partial^2 H / \partial p \partial S$. Differentiating the first Eq. 4.47 with respect to p and the second one with respect to S and making use of this equality, we obtain the Maxwell relation for the enthalpy in the form

$$\left(\frac{\partial T}{\partial p}\right)_S = \left(\frac{\partial V}{\partial S}\right)_p \tag{4.48}$$

The foregoing question about the final volume of a gas after compression from the pressure p to the pressure p' can now be easily answered by computing $V' = (\partial H / \partial p)_S$ for the entropy S and pressure p'. This shows that the enthalpy is a practically useful thermodynamic quantity. The use of the enthalpy is advantageous when it is necessary to analyze isentropic processes with pressure changes. Our conclusions about the enthalpy can be summarized as follows.

Enthalpy *H(S,p)* as a fundamental thermodynamic function. Enthalpy as a function of entropy and pressure is a fundamental thermodynamic function which permits one to compute the volume $V' = (\partial H / \partial p)_S$ and the temperature $T' = (\partial H / \partial S)_p$ after the transition of a system from a state with entropy S and pressure p to a state with the same entropy and pressure p'.

In addition to internal energy and enthalpy, there are at least two other important fundamental thermodynamic functions, namely the Helmholtz function, also called free energy, and the Gibbs function, also called free enthalpy. Why do we have to deal with them? There are two reasons for doing that. First, it is interesting for the sake of mathematical completeness to subject the quantities $U(S, V)$ and $H(S, p)$ to a Legendre transform with respect to the independent variable S. Second, there are practical problems which are difficult to solve using U and H. The reason for this is illustrated in Fig. 4.8c and d. We now turn to the process shown in Fig. 4.8c.

C – Helmholtz Function

In Fig. 4.8c, we symbolically show the isothermal compression of a substance from an initial volume V to a final volume V'. What is the pressure p' after this compression? What is the value of the entropy S' after this compression? Neither $U(S, V)$ nor $H(S, p)$ are suited to answer this question because none of these fundamental thermodynamic functions contains the temperature $T = (\partial U / \partial S)_V$ explicitly which is assumed to be held constant during the process. Let us perform a Legendre transform according to

$$F = U - \left(\frac{\partial U}{\partial S}\right)_V S \tag{4.49}$$

in order to replace entropy as an independent variable by temperature. The so-defined quantity is called Helmholtz function. Taking into account the definition of the temperature, namely $T = (\partial U / \partial S)_V$, the Helmholtz function takes the form

$$F = U - TS \tag{4.50}$$

We can now convince ourselves that the Helmholtz function can indeed be represented as a function of volume and temperature by rewriting the function $T = f_T(S, V)$ symbolically as $S = f_T^{-1}(T, V)$. If we insert this relation into Eq. 4.50, we indeed obtain the desired relation for the Helmholtz function as $F(T, V) = U(f_T^{-1}(T, V), V) - Tf_T^{-1}(T, V)$. The total differential of the Helmholtz function takes the form

$$\mathrm{d}F = \left(\frac{\partial F}{\partial T}\right)_V \mathrm{d}T + \left(\frac{\partial F}{\partial V}\right)_T \mathrm{d}V \tag{4.51}$$

At the same time, Eqs. 4.35 and 4.50 lead to

$$\mathrm{d}F = -S\mathrm{d}T - p\mathrm{d}V \tag{4.52}$$

By comparing the coefficients in front of $\mathrm{d}T$ and $\mathrm{d}V$, we obtain

$$S = -\left(\frac{\partial F}{\partial T}\right)_V, \quad p = -\left(\frac{\partial F}{\partial V}\right)_T \tag{4.53}$$

As with the previous two fundamental thermodynamic functions, the order of differentiation can be changed as $\partial^2 F/\partial T \partial V = \partial^2 F/\partial V \partial T$ from which we derive the Maxwell relation

$$\left(\frac{\partial S}{\partial V}\right)_T = \left(\frac{\partial p}{\partial T}\right)_V \tag{4.54}$$

Our initial question about the pressure of a gas after isothermal compression from V to V' can now be answered by computing $p' = -(\partial F/\partial V)_T$ at the point (T, p'). This demonstrates that the Helmholtz function is not only a mathematically well-defined quantity but a useful quantity as well. The use of the Helmholtz function is advantageous if isothermal processes with variable volume have to be analyzed. Our conclusions can be summarized as follows.

Helmholtz function *F(T, V)* as a fundamental thermodynamic function. The Helmholtz function as a function of temperature and volume is a fundamental thermodynamic function, which permits one to compute the pressure $p' = -(\partial F/\partial V)_T$ and entropy $S' = -(\partial F/\partial T)_V$ of a system which evolves from a state (T, V) into a state (T', V').

D – Gibbs Function

A glance at Fig. 4.8d shows that there remains one open question, namely about the volume V' of a system after an isothermal process which starts at pressure p and ends in a state with pressure p'. In order to solve this problem, we start from enthalpy H and proceed in a similar way as with the transition from internal energy to the Helmholtz function. We perform a Legendre transform from H to a new function G according to

$$G = H - \left(\frac{\partial H}{\partial S}\right)_p S \tag{4.55}$$

thereby replacing entropy as an independent variable by temperature. The new quantity G is called Gibbs function or free enthalpy. Taking into account $T = (\partial H/\partial S)_p$, the new quantity can be written in the form

$$G = H - TS \tag{4.56}$$

It is straightforward to verify that the Gibbs function can be represented as a function of pressure and temperature by rewriting the function $T = f_T(S, p)$ symbolically as $S = f_T{}^{-1}(T, p)$. This is inserted into Eq. 4.56 and leads to the desired expression in the form $G(T, p) = H(f_T{}^{-1}(T, p), p) - T f_T{}^{-1}(T, p)$. The total differential of the Gibbs function reads

$$\mathrm{d}G = \left(\frac{\partial G}{\partial T}\right)_p \mathrm{d}T + \left(\frac{\partial G}{\partial p}\right)_T \mathrm{d}p \tag{4.57}$$

At the same time Eqs. 4.35 and 4.56 imply

$$dG = -SdT + Vdp \tag{4.58}$$

By comparing the coefficients in front of dT and dp, we obtain

$$S = -\left(\frac{\partial G}{\partial T}\right)_p, \quad V = \left(\frac{\partial G}{\partial p}\right)_T \tag{4.59}$$

For the same reasons as in the previous cases, the order of differentiation can be changed, i.e., $\partial^2 G/\partial T\partial p = \partial^2 G/\partial p\partial T$. Hence we obtain the Maxwell relation

$$\left(\frac{\partial S}{\partial p}\right)_T = -\left(\frac{\partial V}{\partial T}\right)_p \tag{4.60}$$

The problem of computing the volume of a gas after compression from p to p' can now be easily solved by evaluating $V' = -(\partial G/\partial p)_T$ at (T, p'). We have thus shown that the Gibbs function is not only a mathematically well-defined entity but a physically useful quantity as well. The use of this quantity is advantageous when temperature and pressure are given as external parameters or when the temperature changes at constant pressure. The Gibbs function will become particularly useful in the analysis of mixing processes and chemical reactions. More precisely, we will use it extensively when deriving the law of mass action in Sect. 5.6 and in the analysis of distillation processes in Sect. 5.7. Our conclusions about the Gibbs function can be summarized as follows.

Gibbs function $G(T,p)$ as a fundamental thermodynamic function. The Gibbs function as a function of temperature and pressure is a fundamental thermodynamic function which permits one to compute the volume $V' = (\partial G/\partial p)_T$ and the entropy $S' = -(\partial G/\partial T)_p$ of a thermodynamic system after a transition from the state (T, p) to the state (T', p').

The main equations relating to fundamental thermodynamic functions are summarized in Fig. 4.9.

Fig. 4.9.
Overview of fundamental thermodynamic functions. Summary of important definitions and properties of internal energy $U(S, V)$, enthalpy $H(S, p)$, Helmholtz function $F(T, V)$ and Gibbs function $G(T, p)$. The relations in the fourth row of each entry are the Maxwell relations

$U(S,V)$ $U = H - pV$ $dU = TdS - pdV$ $\left(\frac{\partial T}{\partial V}\right)_S = -\left(\frac{\partial p}{\partial S}\right)_V$	$H(S,p)$ $H = U + pV$ $dH = TdS + Vdp$ $\left(\frac{\partial T}{\partial p}\right)_S = \left(\frac{\partial V}{\partial S}\right)_p$
$F(T,V)$ $F = U - TS$ $dF = -SdT - pdV$ $\left(\frac{\partial S}{\partial V}\right)_T = \left(\frac{\partial p}{\partial T}\right)_V$	$G(T,p)$ $G = H - TS$ $dG = -SdT + Vdp$ $\left(\frac{\partial S}{\partial p}\right)_T = -\left(\frac{\partial V}{\partial T}\right)_p$

4.7 Determination of the Entropy of Simple Systems

In principle, Eq. 3.2 gives us the possibility to experimentally determine the entropy of a simple system. However, the procedure expressed by this equation is so abstract that it is not applicable for a practical determination of entropy. Moreover, Lieb-Yngvason machines, which have been used as a didactical tool in Chap. 3, do not exist in reality. Fortunately, neither Eq. 3.2 nor a Lieb-Yngvason machine are necessary to experimentally determine the entropy of a given substance. As we will see shortly, the entropy principle and the quantities derived from entropy permit us to formulate a practical measurement procedure which directly yields the entropy.

In order to determine the entropy of a simple system with one work coordinate, it is necessary to perform two series of measurements on the system. In the first measurement, one has to determine the heat capacity $C_V = (\partial U/\partial T)_V$ as a function of temperature and volume. In the second measurement, one has to evaluate the thermal equation of state $p(T, V)$. The measurement of C_V is accomplished by supplying a small amount of energy ΔU to the system while keeping the volume constant and measuring the temperature increase ΔT. If $\Delta U \ll U$ and $\Delta T \ll T$, the derivative in the foregoing definition of the heat capacity can be approximated by $C_V \approx \Delta U/\Delta T$. For determining the thermal equation of the state, it is necessary to measure the pressure of the system for different temperatures. From $p(T, V)$ the auxiliary quantity $\alpha(T, V) = (\partial p/\partial T)_V$ is determined. The functions $C_V(T, V)$ and $\alpha(T, V)$ form the experimental basis for determining the entropy.

In order to compute S from $C_V(T, V)$ and $\alpha(T, V)$, we start from Eq. 4.35 and write it in the form

$$dS = \frac{1}{T}dU + \frac{p}{T}dV \tag{4.61}$$

Next we compute the total differential of $U(T, V)$ and insert it into this equation. The result reads

$$dS = \frac{1}{T}\left(\frac{\partial U}{\partial T}\right)_V dT + \frac{1}{T}\left[\left(\frac{\partial U}{\partial V}\right)_T + p\right]dV \tag{4.62}$$

In order to integrate this equation, we must express its right-hand side by the experimentally determined functions $C_V(T, V)$ and $\alpha(T, V)$. Whereas we can immediately replace the derivative of the internal energy with respect to temperature by the heat capacity, the term $(\partial U/\partial V)_T$ requires some additional work.

Since Eq. 4.62 represents a total differential, the coefficients in front of the terms dT and dV must obey the relation

$$\frac{\partial}{\partial V}\left\{\frac{1}{T}\left(\frac{\partial U}{\partial T}\right)_V\right\}_T = \frac{\partial}{\partial T}\left\{\frac{1}{T}\left[\left(\frac{\partial U}{\partial V}\right)_T + p\right]\right\}_V \tag{4.63}$$

After evaluating the derivatives, this formula takes the simpler form

$$\left(\frac{\partial U}{\partial V}\right)_T = T\left(\frac{\partial p}{\partial T}\right)_V - p \tag{4.64}$$

We have thus expressed the derivative of the internal energy with respect to volume by measurable quantities. We have almost achieved our goal. To finish our work, we insert Eq. 4.64 into 4.62. Making use of the definition of α, we obtain

$$dS = \frac{1}{T} C_V(T,V)dT + \alpha(T,V)dV \tag{4.65}$$

which we can now integrate. We arbitrarily define S_0 for some reference point (T_0, V_0) and integrate Eq. 4.65 from this reference point to the point (T, V) for which the entropy is desired. Since Eq. 4.65 is a total differential, we can take an arbitrary path of integration. A particularly convenient choice is to integrate first over T while keeping V fixed and then to integrate over V at constant T. The result is

$$S(T,V) = S_0 + \int_{T'=T_0}^{T'=T} \frac{1}{T'} C_V(T',V)dT' + \int_{V'=V_0}^{V'=V} \alpha(T,V')dV' \tag{4.66}$$

With this formula, we have achieved our goal to express the entropy of a system through quantities which can be directly measured.

Chapter 5 Specific Applications

Specific Applications

The entropy $S(T)$ of one kilogram of liquid water is

$$S(10\,°C) = 151\ J\,K^{-1}$$
$$S(20\,°C) = 297\ J\,K^{-1}$$
$$S(30\,°C) = 437\ J\,K^{-1}$$

(See for instance Moran and Shapiro 1995, table A-2. The entropy is normalized in such a way that $S = 0$ for liquid water at the triple point. The given values of entropy are for saturated water and correspond to the vapor pressures $p\,(10\,°C) = 0.01228$ bar, $p\,(20\,°C) = 0.02339$ bar and $p\,(30\,°C) = 0.04246$ bar). If 1 kg of cold water with a temperature of 10 °C and 1 kg of warm water with a temperature of 30 °C are connected thermally, they can undergo an adiabatic process which ends in both kilograms of water having the same temperature of 20 °C. This is so because the entropy of the final state ($297 + 297 = 594$) is higher than that of the initial state ($151 + 437 = 588$). By contrast, the opposite adiabatic process cannot occur because the entropy of the compound system would decrease from 594 J K^{-1} to 588 J K^{-1}. A spontaneous generation of a temperature difference is forbidden by the law of entropy increase.

This simple example shows us that entropy permits us to make accurate predictions whether certain adiabatic processes are possible or not. The example also shows us that we have indeed achieved the goal of finding a quantitative measure for the irreversibility of thermodynamic processes as we have desired in the introduction. Now we need to learn how to apply entropy. In the present chapter, we will analyze a number of specific problems which occur in engineering applications or in nature and whose solution is less obvious than in the foregoing introductory example. To illustrate the all-embracing role of entropy, we will focus on such problems whose solution without entropy is either complicated or even impossible.

5.1 Generation of Energy from the Gulf Stream

The German site of the online encyclopedia Wikipedia in its version of 1 January 2005 contained the following information. "Each second the Gulf Stream transports approximately 50 million cubic meters of water, 30 times more than all rivers in the world combined. The Gulf Stream transports approximately 1.4 petawatts (1.4×10^{15} W) of power which is equivalent to the power production of one million nuclear power plants." Shouldn't it be possible to use this enormous potential of nature to solve the energy problems of mankind? The following example, in which entropy plays a central role, helps us to understand the answer.

A – Formulation of the Problem

Let us first translate our question into a well-defined thermodynamic problem with the help of Fig. 5.1. Consider a sample of water from the Gulf Stream with a mass $m = 1\,000$ kg and a temperature $T_H = 25\,°C$. Assume that this thermodynamic system represents our source of energy and is characterized by its internal energy U_H and entropy S_H. We further assume that the water of the Atlantic Ocean outside the Gulf Stream has a temperature $T_C = 10\,°C$ and that its mass M is so large ($M \gg m$) that its temperature remains unchanged when it exchanges energy with our sample.

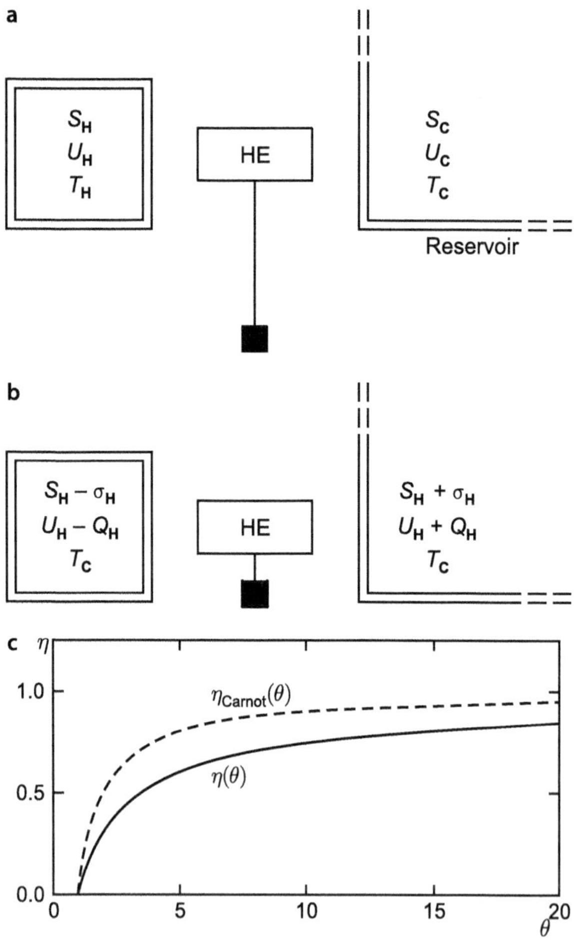

Fig. 5.1. *Conversion of "heat" into work.* (a) One ton of warm water and a large reservoir with cold water are used to generate mechanical energy using a heat engine (HE). In the final state (b), both water reservoirs have the same temperatures. The work performed by the heat engine is symbolically illustrated by lifting a weight. (c) The thermal efficiency η of the process as a function of the dimensionless temperature $\theta = T_H/T_C$ is lower than the thermal efficiency of the Carnot process $\eta_{Carnot} = 1 - T_C/T_H$ for each $\theta > 1$

The warm water of the Gulf Stream possesses an excess internal energy over the cool water of the Atlantic Ocean in the amount of Q_H. This is the energy which the sample would have to release in order to cool down to the temperature of the Atlantic. Assuming a constant specific heat capacity of the water of $c = 4\,184\ \mathrm{J\,kg^{-1}\,K^{-1}}$, we can express the excess internal energy as $Q_H = mc(T_H - T_C)$ and compute its value as $Q_H = 6.28 \times 10^7\ \mathrm{J} = 17.4\ \mathrm{kWh}$. If this energy could be entirely converted into work, one could use it to lift a weight with a mass of 1 000 kg by 6 400 m. As we have shown in Sect. 4.5 (Eq. 4.25), at most a fraction $\eta_{\mathrm{Carnot}} = 1 - T_C / T_H$ of this energy can be converted into work W. In the present example, this amount is approximately equal to 5.03% and corresponds to $W = 0.87$ kWh. We have learned, however, that this figure is only an upper bound. What is the exact value of the maximum work? Is it closer to 5% or to 1%? This question will be addressed next.

We assume that a device extracts the energy Q_H from the warm water, converts part of it into work W and rejects waste heat Q_C into the cold Atlantic. Such a process is called a power cycle. We use the convention that the quantities Q_C, Q_H and W are always positive. The device executing the power cycle is called heat engine. At the end of the power cycle, the previously warm water has cooled down to the temperature T_C and the heat engine has returned into its initial state. As an illustration for the work performed by the heat engine, a lifted weight is included in Fig. 5.1. At the present stage, we will not be interested in the details of the construction of the heat engine. We will return to this question in Sect. 5.4. For the time being, our goal is to compute the maximum work which can be extracted by the heat engine from the compound thermodynamic system consisting of the warm water of the Gulf Stream and the cold water of the Atlantic outside the Gulf Stream.

In the first step, we write down the condition of energy conservation in the form

$$W = Q_H - Q_C \tag{5.1}$$

This equation is not sufficient for a unique determination of W. Indeed, it allows an infinite variety of solutions such as for instance $W = 0$, corresponding to a useless discharge of the available energy into the cold ocean, or $W = Q_H$ which corresponds to conversion of the entire thermal energy into work, which is forbidden by the Kelvin-Planck formulation of the second law of thermodynamics. There are even solutions with $W < 0$, but they do not describe a power cycle because in these cases the weight would have lowered its position in the Earth's gravitational field instead of rising. In order to overcome the non-uniqueness of our problem, we have to consider the balance of entropy in addition to the balance of energy. This requires that we compute the entropy of water.

B – Entropy of an Incompressible Substance

The simplest system for which entropy can be expressed analytically is a substance whose volume is constant and whose heat capacity c is independent of temperature. A material with constant volume is called *incompressible* (Notice that $c_p = c_v$ for an incompressible substance by virtue of the general relation $c_p - c_v = -T(\partial v / \partial T)_p^2 (\partial p / \partial v)_T$ and the condition of constant volume in the form $(\partial v / \partial T)_p = 0$). Since the volume of water remains virtually unchanged in the range of temperatures $10\,°\mathrm{C} \le T \le 25\,°\mathrm{C}$, which we are considering in our present example, we will treat water as incompressible. The thermal equation of state for such a substance has the simple form $V=\text{const}$. In this degenerate case, the volume is no longer a thermodynamic variable and the entire thermodynamic information is contained in the function $S(U)$. Our task is therefore reduced to determining the entropy as a function of the internal energy. Since the internal energy is a unique function of the temperature $T=(\mathrm{d}S/\mathrm{d}U)^{-1}$, we can also use the function $S(T)$ as a basis for all further considerations.

We start our computation of $S(T)$ by assuming that our incompressible substance has a caloric equation of state which is given by

$$U(T)=U_0+mc(T-T_0) \tag{5.2}$$

(see for instance Moran and Shapiro 1995). Here U_0 is the internal energy of the substance at an arbitrary reference temperature T_0. Since internal energy is only determined up to an arbitrary constant, we can set $U_0=0$ without loss of generality. We next compute the differential of the caloric equation of state and obtain

$$\mathrm{d}U=mc\mathrm{d}T \tag{5.3}$$

Comparing this expression with the first $T\mathrm{d}S$-equation ($\mathrm{d}U=T\mathrm{d}S-p\mathrm{d}V$) for the particular case $\mathrm{d}V=0$ leads us to

$$\mathrm{d}U=T\mathrm{d}S \tag{5.4}$$

We now eliminate $\mathrm{d}U$ between Eqs. 5.3 and 5.4 and are led to

$$\mathrm{d}S=\frac{mc}{T}\mathrm{d}T \tag{5.5}$$

This equation is integrated starting from T_0 with the initial condition $S(T_0)=S_0$ and leads us to the desired result

$$S(T)=S_0+mc\ln\frac{T}{T_0} \tag{5.6}$$

which represents the entropy of an incompressible substance.

We could have obtained the same result by inserting the caloric equation of state in the form $C_V=mc$ as well as $\alpha=0$ into Eq. 4.66. Using the caloric equation of state (Eq. 5.2), we can also express the entropy as a function of the internal energy in the form

$$S(U)=S_0+mc\ln\left(1+\frac{U-U_0}{mcT_0}\right) \tag{5.7}$$

As required by the entropy principle, the obtained entropy is a monotonic and concave function of the internal energy, whereas the internal energy is a convex function of the entropy. Equation 5.6 is our first example of an entropy function for a specific thermodynamic system. This auxiliary result can be summarized as follows.

Entropy of an incompressible substance. The entropy of a simple thermodynamic system with constant specific heat capacity and constant volume depends on temperature as $S(T) = S_0 + mc\ln(T/T_0)$. The constants T_0 and S_0 can be chosen arbitrarily, for instance as $T_0 = 273.15$ K and $S_0 = 0$.

Before we return to our original problem, we wish to simplify Eq. 5.7 for the special case $(U - U_0 \ll mcT_0)$. In this case, the internal energy changes so little that the temperature remains virtually constant. We can then invoke the Taylor series expansion $\ln[1 + (U - U_0)/mcT_0] \approx (U - U_0)/mcT_0$ from which we recover the result

$$S - S_0 = \frac{U - U_0}{T_0} \tag{5.8}$$

which is equivalent to Eq. 5.4. In introductory thermodynamics textbooks, a similar relation, namely $dS = \delta Q/T$ is often used for the definition of entropy. Thanks to the Lieb-Yngvason theory, we know now that entropy is not defined through "heat" but rather "heat" through entropy (see also Sect. 4.3).

C – Result and Discussion

Let us now return to our initial question. Our goal is to remove the ambiguity in the determination of the work W by including the entropy balance into our consideration. In order to determine the change of entropy in our compound system consisting of cold and warm water, we first compute the entropy decrease $\sigma_H = S(T_H) - S(T_C)$ of the warm water after it has cooled down. Using Eq. 5.6, we obtain

$$\sigma_H = mc\ln\frac{T_H}{T_C} \tag{5.9}$$

The temperature of the cold water remains virtually unchanged, because its mass is much larger than that of the warm water sample. Hence we can use the Approximation 5.8 for the computation of its entropy increase σ_C. The result is

$$\sigma_C = \frac{Q_C}{T_C} = \frac{Q_H - W}{T_C} \tag{5.10}$$

Taking into account $Q_H = mc(T_H - T_C)$ and remembering that both σ_C and σ_H are positive by definition, we obtain the change of the entropy $\Delta S = \sigma_C - \sigma_H$ of the whole system as

$$\Delta S = mc\left(\frac{T_H - T_C}{T_C} - \ln\frac{T_H}{T_C}\right) - \frac{W}{T_C} \tag{5.11}$$

This equation represents the relation between the work W delivered by the heat engine and the entropy ΔS produced by our compound system. Our process is an adiabatic process, since the sole change outside the compound system is the change in the elevation of a weight. Consequently, we can apply the law of entropy increase in the form $\Delta S \geq 0$ to our process. It follows from this law that the left-hand side of Eq. 5.11 must not become negative. This leads us to the inequality

$$W \leq mc\left((T_{\mathrm{H}} - T_{\mathrm{C}}) - T_{\mathrm{C}} \ln \frac{T_{\mathrm{H}}}{T_{\mathrm{C}}}\right) \tag{5.12}$$

Equations 5.11 and 5.12 display a fundamental link between irreversibility and the efficiency of power cycles. If we "spill" all the thermal energy of the warm water by forming a thermal join with the cold water and let the temperatures equilibrate without extracting any useful work, which corresponds to $W = 0$, then Eq. 5.11 always gives $\Delta S > 0$ (The first term on the right-hand side of Eq. 5.11 is positive. It is left as an exercise to the reader to demonstrate that this is indeed true). Consequently, this will be an irreversible process. By contrast, if we convert a part of the thermal energy of the warm water into work, the entropy production decreases with increasing W, as can be inferred from the negative sign of the last term in Eq. 5.11. When W finally reaches the maximum value which the Inequality 5.12 admits, namely

$$W_{\max} = mc\left((T_{\mathrm{H}} - T_{\mathrm{C}}) - T_{\mathrm{C}} \ln \frac{T_{\mathrm{H}}}{T_{\mathrm{C}}}\right) \tag{5.13}$$

the produced entropy is zero. This limiting case corresponds to a reversible process. With the derivation of Eq. 5.13, we have achieved our goal defined at the beginning of this section. As a by-product, we have obtained a useful statement about the relation between work and irreversibility which can be summarized as follows.

> **The role of irreversibilities in power cycles.** A power cycle between two reservoirs delivers the maximum work if the process is reversible.

Inserting our numerical values into Eq. 5.13, we come to the conclusion that from the available energy in the amount 17.4 kWh a maximum of

$$W_{\max} = 0.45 \text{ kWh} \tag{5.14}$$

can be converted into work, which corresponds to 2.6% only. This work would suffice to lift a body with a mass of 1 000 kg by 160 m rather than by 6 400 m, as mentioned in the beginning of this section.

Our numerical example shows that only a tiny fraction of the energy carried by the Gulf Stream can be converted into work. This, however, does not completely answer our initial question as to why our energy needs cannot be satisfied using this energy source. One could argue that 2.6% of the energy production of one million nuclear power plants is still as much as 26 000 power plants. This would be more than enough to produce all of the energy that we need. However, the investment and operating costs of such a "Gulf Stream economy" based on today's technology are prohibitively high, not to mention the unknown ecological consequences.

D – Suggestions for Further Study

Having completed our analysis, let us look at Eq. 5.13 from a more global perspective. If we replace the dimensional quantity W with the non-dimensional efficiency $\eta = W_{max}/\,mc(T_H - T_C)$ and introduce a non-dimensional temperature defined by $\theta = T_H/\,T_C$, then Eq. 5.13 can be recast in a particularly compact form as

$$\eta = 1 - \frac{\ln\theta}{\theta - 1} \tag{5.15}$$

The function $\eta(\theta)$, which is plotted in Fig. 5.1c, describes the maximum fraction of the thermal energy that can be converted into work. The quantity W as defined by Eq. 5.13 is also called *exergy*. Observe that for $\theta > 1$ the efficiency η is always lower than the Carnot efficiency $\eta_{Carnot} = 1 - T_C/\,T_H$ whose non-dimensional form is

$$\eta_{Carnot} = \frac{\theta - 1}{\theta} \tag{5.16}$$

We leave it as an exercise to the reader to demonstrate that this is indeed true. The relation $\eta < \eta_{Carnot}$ is a consequence of the fact that the source of the thermal energy cools down in the course of the process, whereas the Carnot efficiency holds for a process where the temperature of the hot reservoir is constant.

To deepen our understanding of exergy, it is instructive to analyze the limiting cases $\theta \to 1$ and $\theta \to \infty$ in more detail. In the first case, which corresponds to a vanishingly small temperature difference between the heat source and the reservoir, Eq. 5.15 takes the form $\eta \approx (\theta - 1)\,/\,2$. We can verify this by introducing the non-dimensional variable $x = \theta - 1$ and performing a Taylor expansion of Eq. 5.15 around $x = 0$ for $x \ll 1$ using $\ln(1 + x) \approx x - x^2/\,2$. Translating the result back into dimensional quantities, we obtain

$$W_{max} = mc\frac{(T_H - T_C)^2}{2T_C} \tag{5.17}$$

This equation shows that the exergy has a quadratic minimum at $T_H = T_C$. This is the mathematical expression of the difficulty to harvest energy from small temperature differences such as those occurring in nature.

The opposite limit, namely $\theta \to \infty$, corresponds to the case of very large temperature differences. In this case, $\eta \to 1$. In dimensional variables, the result reads

$$W_{max} = mc(T_H - T_C) \tag{5.18}$$

This equation tells us that for large temperature differences, most of the thermal energy can be converted into work. This is in agreement with the observation that the Carnot efficiency tends to unity for $\theta \to \infty$.

Finally, we would like to briefly address the case $T_H < T_C$ $(\theta < 1)$ in which the energy source has a lower temperature than the reservoir. In this case, the curve plotted in

Fig. 5.1c tends to $-\infty$ and seems to violate the principle according to which the efficiency is always positive and its absolute value is smaller than one. This paradox is resolved if we convince ourselves that $\eta < 0$ because $Q_H < 0$, i.e., the "energy source" absorbs energy rather than delivers energy. The fact that the absolute value of the efficiency is larger than one is due to the fact that for $\theta < 1$ the quantity η should no longer be interpreted as an efficiency because Q_H and Q_C change their roles and Q_C plays the role of the energy source. Nevertheless, it is interesting to note that for $\theta < 1$ the obtained work is higher than the energy necessary to compensate the temperature difference. For instance, for $T_H = 1$ K and $T_C = 283.15$ K we obtain $\eta = -4.67$. This implies that using an ice cube with a temperature of -272.15 °C can produce 4.67 times more work than the thermal energy which would be necessary to bring the ice cube to the temperature of 10 °C.

5.2 Air Conditioning

If entropy and the entropy principle didn't exist, devices with remarkable properties could be constructed. For instance, it would be possible to build a car which does not only run without fuel but which additionally provides air conditioning in the summer at no expense. Each second its engine would suck in 1 kg or approximately one cubic meter of air with a temperature of, say, 40 °C and extract 30 kJ of energy from it. This energy would be transformed into work and could be used to drive the car with a power of 30 kW. In addition, the air would cool down to 10 °C and could be used for air conditioning. Such a car would need no fuel and no additional air conditioning system.

Unfortunately, the construction of such cars is impossible because it would violate the second law of thermodynamics in the formulation of Kelvin and Planck. Indeed, such a car would do nothing more than cool a system, namely the parcel of hot air, and transform this energy entirely into work. This is impossible, as we have discussed in Sect. 4.4 B. Therefore, the designers of cars are left with no other choice than to develop engines with the lowest possible energy consumption and devise air conditioning systems which consume the smallest possible part of the valuable power delivered by the engine. Before designing an air conditioning system, however, it is important to know how much work has to be spent in order to cool a given amount of air to a certain temperature. The analysis of this question will occupy the remainder of this section.

A - Formulation of the Problem

We start our consideration by translating our problem into the language of thermodynamics using Fig. 5.2. Let us consider 1 kg of hot air with a temperature $T_H = 40$ °C and atmospheric pressure $p = 1.013$ bar. We assume that the air conditioning system denoted *AC* in Fig. 5.2 extracts the energy Q_C from the air. As a result, the air cools down to $T_C = 10$ °C. In order to accomplish this task, we must deliver the work W to the air conditioning system which is provided by an external energy source. This source is symbolically represented as a weight that has lowered its position in the gravitational field of the Earth. Finally the waste heat Q_H is rejected to the ambient air which we consider to be a huge reservoir with constant temperature. We use the convention that the quantities Q_C, Q_H and W are always positive. For the time being, we do not care about the details of the construction of the air conditioning system. We will extensively discuss refrigeration systems in Sect. 5.5. We only assume that at the end of the process, the air conditioning system returns to its initial state.

The energy balance for the process sketched in Fig. 5.2 reads

$$W = Q_H - Q_C \tag{5.19}$$

When interpreting the quantities, we have to take in to account a small but important difference to the problem discussed in Fig. 5.1. In the previous example, we considered the working substance – water – as incompressible. In the present problem, the

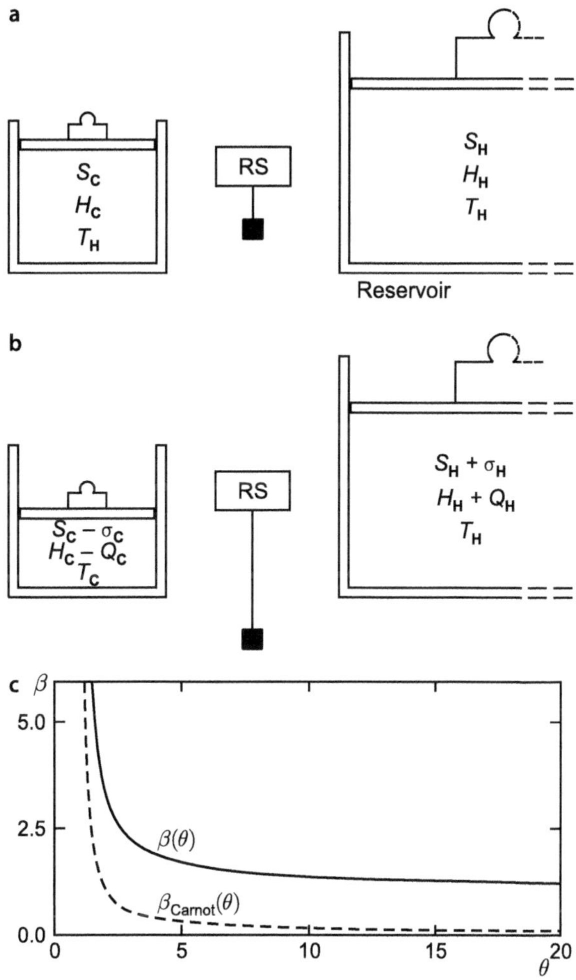

Fig. 5.2. *Conversion of work into "cold".* (a) One kilogram of hot air is cooled using a refrigeration system (RS) and a large reservoir of hot air. In the final state (b) the sample of air has a lower temperature than the reservoir. The work transferred to the refrigeration system is symbolically illustrated by lowering a weight. (c) The coefficient of performance β of the process as a function of the dimensionless temperature $\theta = T_H / T_C$ is higher (not lower!) than the coefficient of performance of the Carnot refrigeration process $\beta_{Carnot} = T_C / (T_H - T_C)$ for each $\theta > 1$. This paradox is resolved in the text

working substance is air. For the temperature difference $\Delta T = 30$ K considered here, the volume of the air changes by about 10%. It is appropriate to take this change of volume into account when formulating the energy balance for our problem. In contrast to the previous example, Q_C is no longer equal to the change of the internal energy ΔU of the cold air. Indeed, the weight which symbolizes the constant pressure in Fig. 5.2a lowers its position slightly during the cooling process. Hence the energy extracted from the cold air is equal to the sum of the change of internal energy ΔU and the change $p\Delta V$ in the potential energy of the weight keeping the pressure of the cold air constant. This is expressed by the energy balance $Q_C = \Delta U + p\Delta V$ (both ΔU and ΔV are assumed to be positive). The right-hand side of this relation is equal to the change of the enthalpy $H + pV$. We have thus shown that the energy balance can be written in a compact form as $Q_C = \Delta H$. It is for this reason that we have used enthalpy rather than internal energy to characterize the states in Fig. 5.2a and b.

As in the previous example, the energy balance (Eq. 5.19) alone is not sufficient to determine W. For a given value of Q_C, the equation admits solutions with any value of W with the property $W \geq 0$ including the case $W = 0$. However, this case would violate the second law of thermodynamics since it would imply the possibility of cooling the air without doing any work. In order to solve the problem uniquely, we have to include the entropy into our consideration. This requires that we compute the entropy of air.

B – Entropy of an Ideal Gas

The simplest thermodynamic model for the behavior of a gas is the ideal gas model. An ideal gas is described by the thermal equation of state $pV = n\mathrm{R}T$ and the caloric equation of state

$$H(T) = H_0 + nc_p(T - T_0) \tag{5.20}$$

(Moran and Shapiro 1995). Here n denotes the number of moles of the gas with the unit mol, $\mathrm{R} = 8.314\ \mathrm{J\,mol^{-1}\,K^{-1}}$ is the universal gas constant and c_p the molar heat capacity at constant pressure which we assume to be independent of temperature. Let H_0 denote the enthalpy of the gas at a reference temperature T_0. Since only differences in enthalpy have a physical meaning, we can assign an arbitrary value to the enthalpy constant. A particularly convenient choice is $H_0 = 0$. We now form the differential

$$\mathrm{d}H = nc_p\mathrm{d}T \tag{5.21}$$

of the caloric equation of state and compare it with the second $T\mathrm{d}S$-equation (Eq. 4.46)

$$\mathrm{d}H = T\mathrm{d}S + V\mathrm{d}p \tag{5.22}$$

We then eliminate $\mathrm{d}H$ between the last two equations and obtain

$$\mathrm{d}S = n\left(\frac{c_p}{T}\mathrm{d}T - \frac{\mathrm{R}}{p}\mathrm{d}p\right) \tag{5.23}$$

Now we integrate this equation starting from T_0 and p_0 using the initial condition $S(T_0, p_0) = S_0$. As result, we obtain the desired equation for the entropy of the ideal gas in the form

$$S(T,p) = S_0 + n\left(c_p \ln\frac{T}{T_0} - \mathrm{R}\ln\frac{p}{p_0}\right) \tag{5.24}$$

To write the entropy as a fundamental thermodynamic function $S(H, p)$, we recast the caloric equation of state into an equation for the temperature and insert the result into the foregoing equation. The result is

$$S(H,p) = S_0 + n\left[c_p \ln\left(1 + \frac{H - H_0}{nc_pT_0}\right) - \mathrm{R}\ln\frac{p}{p_0}\right] \tag{5.25}$$

This equation is our second entropy function for a specific thermodynamic system. We summarize the result as follows.

Entropy of an ideal gas. The entropy of an ideal gas with constant molar heat capacity c_p as a function of temperature and pressure is

$$S(T,p) = S_0 + n[c_p \ln(T/T_0) - \mathrm{R}\ln(p/p_0)]$$

The constants T_0, p_0 and S_0 can be chosen arbitrarily, for instance as $T_0 = 273.15$ K, $p_0 = 1.013$ bar and $S_0 = 0$.

Before we return to our problem, we wish to evaluate Eq. 5.25 for the particular case $H - H_0 \ll nc_pT_0$ which corresponds to a very small enthalpy change. By expanding Eq. 5.25 into a Taylor series and remembering that $\ln(1 + x) \approx x$ for $x \ll 1$, we recover the result

$$S - S_0 = \frac{H - H_0}{T_0} \tag{5.26}$$

which is equivalent to the second $T\mathrm{d}S$-equation for constant pressure ($\mathrm{d}p = 0$) which takes the form $\mathrm{d}S = \mathrm{d}H/T$.

C - Result and Discussion

With Eq. 5.24, we can compute the entropy change ΔS in the compound system consisting of cold and hot air when it undergoes the process sketched in Fig. 5.2. Using Eq. 5.24, we compute the entropy decrease $\sigma_\mathrm{C} = S(T_\mathrm{H}, p_0) - S(T_\mathrm{C}, p_0)$ in the cold air as

$$\sigma_\mathrm{C} = nc_p \ln\frac{T_\mathrm{H}}{T_\mathrm{C}} \tag{5.27}$$

Since the reservoir is huge, its enthalpy and entropy changes are very small in comparison with its total enthalpy and entropy. Hence, we can compute the entropy increase σ_H using the approximate formula (Eq. 5.26). The result is

$$\sigma_H = \frac{Q_H}{T_H} = \frac{Q_C + W}{T_H} \tag{5.28}$$

The entropy increase of the compound system $\Delta S = \sigma_H - \sigma_C$ is

$$\Delta S = nc_p \left(\frac{T_H - T_C}{T_H} - \ln \frac{T_H}{T_C} \right) + \frac{W}{T_H} \tag{5.29}$$

For the same reasons as in Sect. 5.1, our process is adiabatic and must satisfy the condition $\Delta S \geq 0$ imposed by the law of entropy increase. Applying this relation to Eq. 5.29, we obtain

$$W \geq nc_p \left[T_H \ln \frac{T_H}{T_C} - (T_H - T_C) \right] \tag{5.30}$$

The right-hand side of this inequality is always positive and cannot become zero. Hence the left-hand side cannot become zero either, and it is therefore impossible to cool air without doing any work ($W = 0$). As in Sect. 5.1, Eq. 5.29 reflects an important connection between the efficiency of a refrigeration process and irreversibilities. The smaller the produced entropy (ΔS on the left-hand side of Eq. 5.29) and the "weaker" the irreversibility, the less work is necessary for cooling (W on the right-hand side of Eq. 5.29) and the more efficient the refrigeration process is. The law of entropy increase places a limit upon the increase of efficiency. This limit is reached if we use the equality sign in the Relation 5.30. We then obtain the minimum work in the form

$$W_{min} = nc_p \left[T_H \ln \frac{T_H}{T_C} - (T_H - T_C) \right] \tag{5.31}$$

This corresponds to the case $\Delta S = 0$, which characterizes a reversible process. We can summarize our conclusions as follows.

> **Role of irreversibilities in refrigeration cycles.** A refrigeration process between two reservoirs requires the lowest possible amount of work if the exchange of energy takes place in a reversible manner.

In order to compute the numerical value of W_{min} for our example, we use $nc_p = 1.005 \times 10^3$ J K^{-1} for the heat capacity of 1 kg of air. Taking $T_H = 313.15$ K and $T_C = 283.15$ K in Eq. 5.31 we obtain

$$W_{min} = 1563 \text{ J} \tag{5.32}$$

This is the answer to our question.

To develop an appreciation for this numerical value, we note that with the same amount of energy a mass of 1 kg can be lifted to a height of approximately 160 m. It is remarkable that so much work must be performed to cool air by only 30 degrees. However, it is also remarkable that the work is very small in comparison with the enthalpy decrease $Q_C = nc_p(T_H - T_C) = 30.15$ kJ of the cold air. This energy corresponds to lifting 1 kg to a height of 3 100 m. The ratio of the enthalpy decrease to the performed work is thus equal to $\beta = 19.5$ and expresses that we need to use only one Joule of work to extract 19.5 Joule of enthalpy from the air.

D – Suggestions for Further Study

After having answered our initial question, let us have a look at our main result, Eq. 5.31, from a more general perspective. If we replace W_{min} by the non-dimensional coefficient of performance $\beta = Q_C / W_{min}$ and use the abbreviation $\theta = T_H / T_C$, Eq. 5.31 takes the compact form

$$\beta = \left(\frac{\theta \ln \theta}{\theta - 1} - 1 \right)^{-1} \tag{5.33}$$

The function $\beta(\theta)$ is shown in Fig. 5.2c. A small temperature difference which was a curse in power cycles turns into a blessing when considering refrigeration cycles. We can see from Fig. 5.2c that the coefficient of performance increases when the temperature difference becomes smaller.

The limits $\theta \to 1$ and $\theta \to \infty$ deserve particular attention. In the former case, which corresponds to a small temperature difference, Eq. 5.33 takes the form $\beta \approx (\theta - 1)^{-1}$. We can verify this by introducing the new variable $x = \theta - 1$ and expanding Eq. 5.33 into a Taylor series around $x = 0$ taking into account the relation $\ln(1 + x) \approx x$ holding for $x \ll 1$. If we translate the resulting equation back to dimensional variables, we obtain

$$W_{min} = nc_p \frac{(T_H - T_C)^2}{2T_C} \tag{5.34}$$

This equation shows us that the work necessary for cooling down a quantity of air from an initial temperature T_H to a final T_C has a quadratic minimum at $T_H = T_C$. This is the mathematical expression for the fact that it is easy to create small temperature differences.

In the opposite limit $\theta \to \infty$, which corresponds to the case of very large temperature differences, we obtain $\beta \approx 1 / \ln \theta$. Written in dimensional variables, this reads

$$W_{min} = nc_p(T_H - T_C) \ln \frac{T_H}{T_C} \tag{5.35}$$

This result shows us that the creation of large temperature differences requires much more work than the enthalpy $nc_p(T_H - T_C)$ which has to be extracted from the system.

At this point, we would like to mention a paradox. The coefficient of performance in our example is $\beta = 19.5$. If we compute the Carnot coefficient of performance $\beta_{Carnot} = T_C / (T_H - T_C)$, whose non-dimensional form is

$$\beta = (\theta - 1)^{-1} \tag{5.36}$$

we obtain a value of only $\beta = 9.43$! How can it be possible that the Carnot coefficient of performance is smaller than the real coefficient of performance? Intuitively we would expect that the coefficient of performance of a process cannot exceed the Carnot coefficient of performance. The paradox can be resolved if we recall under which assumptions we have performed our derivation in Sect. 4.5. Equation 5.36 is only valid under the assumption that T_H and T_C are the temperatures *before* the cooling process. In the present case, however, both temperatures are equal at the beginning of the process. Hence we cannot directly compare the Eqs. 5.36 and 5.33.

Nevertheless, our observation $\beta(\theta) > \beta_{Carnot}(\theta)$ can be qualitatively understood if we remember that the heat transfer in our example takes place at temperature differences smaller than $T_H - T_C$.

As already mentioned in Sect. 4.5, each refrigeration system can in principle also be used as a heat pump. With the method developed here, we can also answer the question how much work a heat pump must perform in order to heat 1 kg of air from an initial temperature of $T_C = 10\,°C$ to a final temperature of $T_C = 40\,°C$. Assuming that the ambient temperature remains at $T_C = 10\,°C$, we can solve this problem through a slight modification of our foregoing calculation. The obtained minimum work is

$$W_{min} = nc_p\left[(T_H - T_C) - T_C \ln\frac{T_H}{T_C}\right] \tag{5.37}$$

and the maximum possible coefficient of performance $\beta = Q_H / W_{min}$ is given by

$$\beta = \left(1 - \frac{\ln\theta}{\theta - 1}\right)^{-1} \tag{5.38}$$

Up to now, we have always assumed that the specific heat capacity of air is independent of temperature. This assumption is no longer justified if the temperature differences become large. In general, the internal energy and enthalpy of an ideal gas are given by the relations $dU = nc_V(T)dT$ and $dH = nc_p(T)dT$, respectively. Here the specific heat capacities at constant volume and constant pressure are related by $c_p(T) - c_V(T) = R$. A short computation, which we recommend as an exercise, leads to the result that in this general case the entropy $S(T, p)$ has the form

$$S(T,p) = S_0 + n\left\{\int_{T'=T_0}^{T'=T} \frac{c_p(T')}{T'} dT' - R\ln\frac{p}{p_0}\right\} \tag{5.39}$$

whereas $S(T, V)$ is given by

$$S(T,V) = S_0 + n\left\{\int_{T'=T_0}^{T'=T} \frac{c_V(T')}{T'} dT' + R\ln\frac{V}{V_0}\right\} \tag{5.40}$$

(see for instance Moran and Shapiro 1995).

5.3 Ice Skating

Why is it possible to skate on ice but not on an equally smooth glass plate? The answer to this question is closely related to the pressure dependence of the melting temperature of ice. Under atmospheric pressure, ice melts at 0 °C. If the pressure exceeds atmospheric pressure, the melting point of ice decreases. The pressure at the point where the blade is in contact with the ice is very high. In this small region the ice melts and forms a thin liquid film which enables the skater to slide with very little friction. How strong is the depression of the melting point of ice under the influence of high pressure? Has the melting point depression anything to do with the fact that water increases its volume upon freezing and that the high pressure "pushes" the system back into the liquid state? These are the questions which we will address below.

A – Formulation of the Problem

Let us first translate our question in a well-defined thermodynamic problem referring to Fig. 5.3. We consider 1 kg of water whose temperature is changed while the pressure is kept constant.

As shown in Fig. 5.3, the melting temperature at atmospheric pressure ($p = 1.013$ bar) is $T_0 = 0$ °C. It is known from experiments that the melting temperature of water decreases with increasing pressure. Our goal is to compute the melting temperature as a function of pressure. In particular, we are interested in the question whether it is possible to perform such a computation on the basis of data at *atmospheric pressure* only. Moreover, we would like to know whether the decrease of the solidification temperature with pressure has anything to do with the fact that water expands upon freezing. As we will see, the results to be derived in this section are not only valid for the coexistence between the solid and liquid phases of a substance but also for the transition between liquid and gaseous state – boiling – and the transition from solid to gaseous state – sublimation. In any case, the knowledge of entropy is an indispensible prerequisite for achieving our goal.

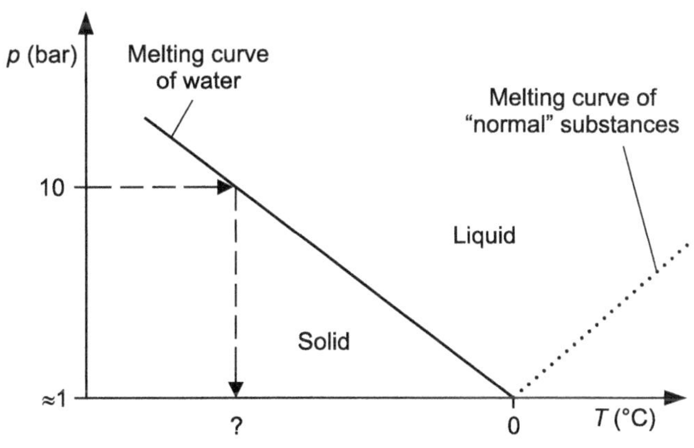

Fig. 5.3. *Melting of ice under pressure.* Schematic of a part of the phase diagram of water which shows the melting point depression under the influence of pressure. In contrast to water, the melting temperature of many other substances increases with pressure. This is indicated by the dotted line which originates from $T = 0$ °C

B – Entropy of a Two-Phase System

Our system is a chemically uniform substance which simultaneously exists in solid and liquid states. Such a system is called a two-phase system. To determine its entropy we consider the hypothetical melting experiment shown in Fig. 5.4a. Since the experiment takes place at constant pressure, it is convenient to use the enthalpy instead of the internal energy to characterize our system. The experiment consists of heating a given amount of ice by increasing its enthalpy H in small steps and measuring the temperature of the system. As long as the whole system is in the solid state, i.e., $H < H_1$, the temperature of the ice is a monotonically increasing function of its enthalpy. As soon as the enthalpy has reached the value $H = H_1$ denoted by A in Fig. 5.4a, further increase of the enthalpy does not lead to an increase of the temperature. We rather observe that an increasing fraction of the ice melts and the volume the of system decreases. One representative point from this family of equilibrium states is marked with B in Fig. 5.4a. Once the enthalpy has reached the value $H = H_2$ corresponding to state C, all the water will be in the liquid state. Now further increase of the enthalpy will again lead to an increase of the temperature and volume. In what follows, we shall exclusively consider the case $H_1 < H < H_2$ where we have a two-phase system.

Figure 5.4b reflects an important property of two-phase systems. Once the pressure is given, the temperature is no longer an independent quantity and adjusts itself to the value of the pressure. To characterize the thermodynamic system uniquely, it is necessary to introduce a second independent thermodynamic variable. We can either use the enthalpy itself or the liquid fraction $x = (H - H_1) / (H_2 - H_1)$ which describes the

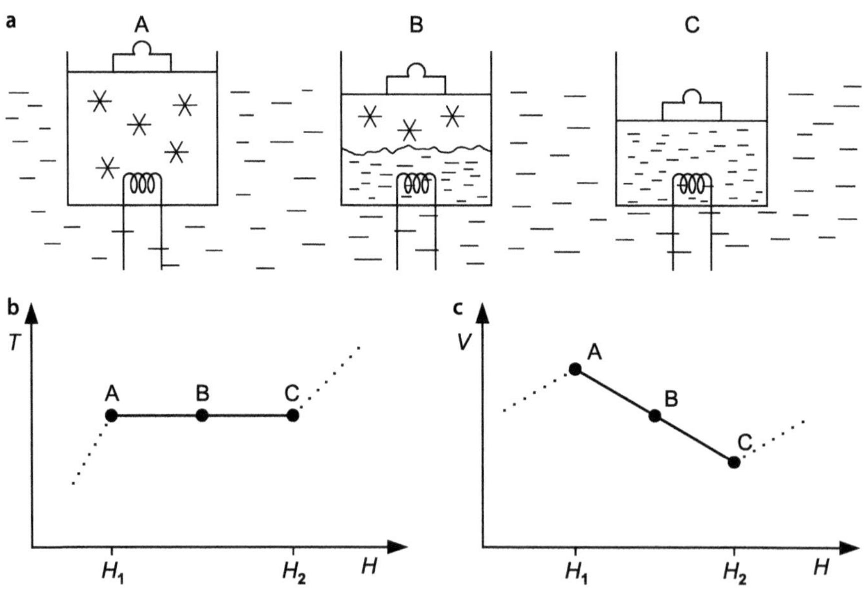

Fig. 5.4. *Experimental basis for the determination of the entropy of a water-ice mixture.* (a) Melting of ice at constant pressure using an electric heater. (b) During the melting process, the temperature remains constant. (c) The volume of the system decreases upon melting. The volume change is not to scale

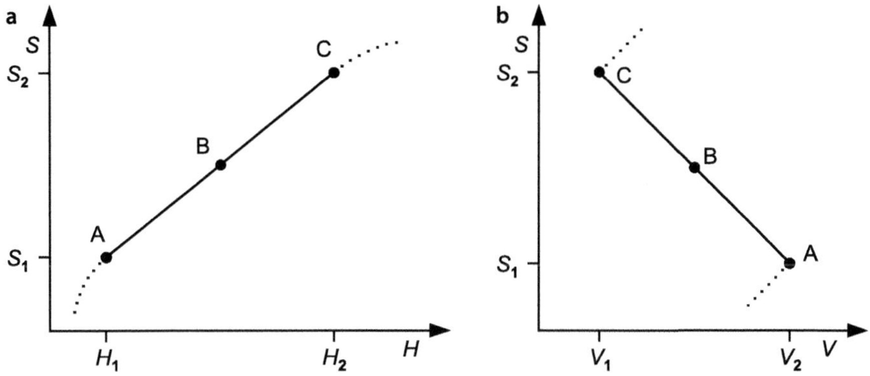

Fig. 5.5. *Entropy of a water-ice mixture.* Schematic representation of the entropy (a) as a function of the enthalpy, (b) as a function of the volume. In both cases, pressure and temperature are constant. Notice that $S(V,p)$ is a multiple-valued function of the volume because the volume of water increases upon freezing. However, the function $S(U,V)$ is single-valued as required by the entropy principle. The symbols A, B, and C refer to the states in Fig. 5.4a

mass of the liquid phase relative to the total mass of the system. When considering phase transitions between liquid and gaseous states, the analogous quantity is called *quality of vapor.*

Since the pressure in Fig. 5.4 is constant, the second $T\mathrm{d}S$-equation (Eq. 4.46) takes the simple form $\mathrm{d}H = T\mathrm{d}S$. The temperature can be represented as $T = (\partial U/\partial S)_V$ or as $T = (\partial H/\partial S)_p$ and is constant in the two-phase system, as we have seen earlier. Hence the entropy is a linear function of the enthalpy. This circumstance is illustrated in Fig. 5.5a. Denoting the entropy for $H = H_1$ as S_1, an integration of the foregoing differential with the initial condition $S(H_1) = S_1$ yields

$$S(H,p) = S_1 + \frac{H - H_1}{T} \tag{5.41}$$

Observe that the quantities S_1, H_1, and T are functions of the pressure which can be determined experimentally. $H_2 - H_1$ is called the *latent heat of melting.* We can obtain an alternative representation of entropy if we replace enthalpy by x in Eq. 5.41 and write the entropy of liquid water as $S_2 = S_1 + (H_2 - H_1)/T$. A straightforward computation leads to

$$S(x,p) = (1 - x)S_1 - xS_2 \tag{5.42}$$

This relation has a simple physical interpretation. The entropy of a two-phase system is the weighted average of the entropies of the pure solid and the pure liquid. The functions $S_1(p)$ and $S_2(p)$ are tabulated, for instance in table A-2 of the textbook Moran and Shapiro (1995). An important consequence of the entropy expression (Eq. 5.41) is that that the Gibbs function $G = H - TS$ is constant for a two-phase system at constant pressure. Equation 5.41 is the third example of an expression of entropy for a specific thermodynamic system. We summarize this result as follows.

Entropy of a two-phase system. The entropy of a two-phase system is a linearly increasing function of enthalpy and has the form $S(H,p) = S_1 + (H - H_1)/T$. The quantities $S_1(p)$, $H_1(p)$, and $T(p)$ can be determined experimentally and are usually given in the form of tables.

C – Result and Discussion

How can the knowledge of entropy help us in determining the melting temperature? The entropy is the piece of information which connects the unknown melting temperature with the latent heat of melting, which can be measured conveniently. The second important piece of information is the Maxwell relation $(\partial S/\partial V)_T = (\partial p/\partial T)_V$ (Eq. 4.48) which we have derived in Sect. 4.6 B. By virtue of the linear relation between entropy and volume, illustrated in Fig. 5.5, the left-hand side of this equation can be written as

$$\left(\frac{\partial S}{\partial V}\right)_T = \frac{S_2 - S_1}{V_2 - V_1} \tag{5.43}$$

Since the pressure depends only on temperature, we can replace the partial derivative on the right-hand side of the Maxwell relation with the ordinary derivative

$$\left(\frac{\partial p}{\partial T}\right)_V = \frac{\mathrm{d}p}{\mathrm{d}T} \tag{5.44}$$

The Maxwell relation then takes the form

$$S_2 - S_1 = \frac{\mathrm{d}p}{\mathrm{d}T}(V_2 - V_1) \tag{5.45}$$

At the same time we can write (Eq. 5.41) for $H = H_2$ as

$$S_2 - S_1 = \frac{1}{T}(H_2 - H_1) \tag{5.46}$$

We now eliminate the entropy difference between the last two equations and obtain the result

$$\frac{\mathrm{d}p}{\mathrm{d}T} = \frac{1}{T}\frac{H_2 - H_1}{V_2 - V_1} \tag{5.47}$$

This is the *Clapeyron equation* which contains the answers to all the questions formulated in the beginning of this section. This equation connects the slope $\mathrm{d}p/\mathrm{d}T$ of the melting curve $p(T)$, shown in Fig. 5.3, to the differences of enthalpy and volume between the solid and liquid phases, respectively.

Two properties of this equation deserve particular attention. First, the slope of the melting curve has the same sign as the difference $V_2 - V_1$ between the volumes of the liquid and solid phases. This is due to the fact that $H_2 - H_1$ is always positive. Since water expands upon freezing, we have $V_2 - V_1 < 0$ from which we can conclude that $p(T)$ has negative slope as shown in Fig. 5.3. This implies that the melting point decreases with increasing pressure, as already noticed. Second, the Clapeyron equation connects enthalpies and volumes at a given pressure with the melting curve at adjacent pressures. If the curvature of $p(T)$ is small, we can linearize $p(T)$ near the point (T_0, p_0) as follows

$$p(T) \approx p_0 + \left(\frac{\mathrm{d}p}{\mathrm{d}T}\right)(T - T_0) \tag{5.48}$$

or

$$T(p) \approx T_0 + \left(\frac{\mathrm{d}p}{\mathrm{d}T}\right)^{-1} (p - p_0) \tag{5.49}$$

This equation permits us to compute the melting temperature of water at elevated pressures on the basis of data at atmospheric pressure only. For instance, to approximately compute the melting temperature of ice at 100 bar we only need to know the latent heat of melting and the volume change at atmospheric pressure. This is an example of the far-reaching consequences of the entropy principle.

To illustrate the utility of the Clapeyron equation, let us calculate the melting point of ice at high pressure. Using $H_1 = -333$ kJ, $H_2 = 0$, $V_1 = 1.0908 \times 10^{-3}$ m^3, $V_2 = 1.0002 \times 10^{-3}$ m^3 we obtain

$$\frac{\mathrm{d}p}{\mathrm{d}T} = -135 \text{ bar K}^{-1} \tag{5.50}$$

Assuming that $p(T)$ of ice can be linearized in the vicinity of $p = p_0$ and using Eq. 5.49 for 10 bar we obtain a melting temperature of

$$T \approx -0.074\,°\mathrm{C} \tag{5.51}$$

This result formally concludes our analysis. Unfortunately, the result does not provide an answer to our initial question as to why ice skating is possible. Indeed, the melting point depression predicted by our theory is not very large. Even if we are very generous and assume a pressure of 1 000 bar under the skates of an ice skater, which would correspond to a person with 100 kg mass acting upon an area of 10 mm^2, the obtained melting temperature would still be as "high" as −7.4 °C. If our initial hypothesis about the thin liquid film were true, ice skating would be impossible below −7.4 °C. However, it is known that ice skating is possible even for people lighter than 100 kg and even at temperatures much below −7.4 °C. Indeed ice skating is possible down to temperatures of around −30 °C. Where is the flaw in our theory?

The physics of ice skating and also of skiing are much more complex than we have assumed up to now. The sliding properties of skates are not only determined by the pressure dependence of the melting temperature of water but also by frictional heat as well as the particular properties of the surface of ice. The latter have not been understood until recently, as can be seen for instance in the scientific article by Engemann et al. (2004). A popular presentation of this fascinating subject, involving thermodynamics, fluid dynamics and solid-state physics can be found in an article by Rosenberg (2005).

D – Suggestions for Further Study

The applicability of the Clapeyron equation is not restricted to phase transitions between solid and liquid states. We can illustrate this using a numerical example for the boiling point elevation of water. At atmospheric pressure, the boiling point of water is $T = 100\,°C$. The enthalpy of evaporation of 1 kg of water is $H_2 - H_1 = 2\,257$ kJ, and the volume change is $V_2 - V_1 = 1.676\ m^3$. Applying the Clapeyron equation we obtain

$$\frac{dp}{dT} = +0.036\ \text{bar}\ K^{-1} \tag{5.52}$$

for the slope of the function $p(T)$. For a system consisting of a gaseous and a liquid phase, this function is also called the vapor pressure curve. Applying the linearized Eq. 5.49, we come to the conclusion that an overpressure (underpressure) of 0.036 bar depresses (increases) the boiling point of water by one degree. At the top of Mount Everest, the air pressure is only one fourth of the atmospheric pressure at sea level, or $p \approx 0.25$ bar. According to Eq. 5.49, water would boil there already at $T \approx 80\,°C$. As a result, the time for the preparation of food can increase significantly. This is the reason why alpinists cannot enjoy such fine food as Thuringian Dumplings (see *http://de.wikipedia.org/wiki/Thüringer_Klöße*) but must have recourse to tasteless industrial food. In contrast, in a pressure cooker the pressure is approximately 2 bar. This leads to an increase of the boiling point to $T \approx 127\,°C$ and reduces the time for cooking.

As a suggestion for further study, let us finally mention the particular case in which phase "2" is a gas which is described by the equation of state of an ideal gas $pV_2 = nRT$ and which has a much higher volume than phase "1", i.e., $V_2 \gg V_1$. This case can occur either for boiling or sublimation. In this case, we can neglect V_1 in the denominator of Eq. 5.47 and express V_2 by the equation of state of the ideal gas. In addition, we write $H_2 - H_1 = n\Delta h$, move the pressure to the left-hand side and use $p^{-1}dp = d(\ln p)$. This transforms the Clapeyron equation into

$$\frac{d\ln p}{dT} = \frac{\Delta h}{RT^2} \tag{5.53}$$

This is the *Clausius-Clapeyron equation*, which permits us to compute the molar evaporation enthalpy Δh from the slope of $p(T)$ or vice versa. Our result is an interesting consequence of the entropy principle because it allows us to extrapolate the behavior of a system at high pressure from measurements at atmospheric pressure. If, in addition, the evaporation enthalpy depends only weakly on pressure, we can regard Δh as a constant and integrate the Clausius-Clapeyron equation using the initial condition $p(T_0) = p_0$. The result is

$$p(T) = p_0 \frac{\exp(\Delta h/RT)}{\exp(\Delta h/RT_0)} \tag{5.54}$$

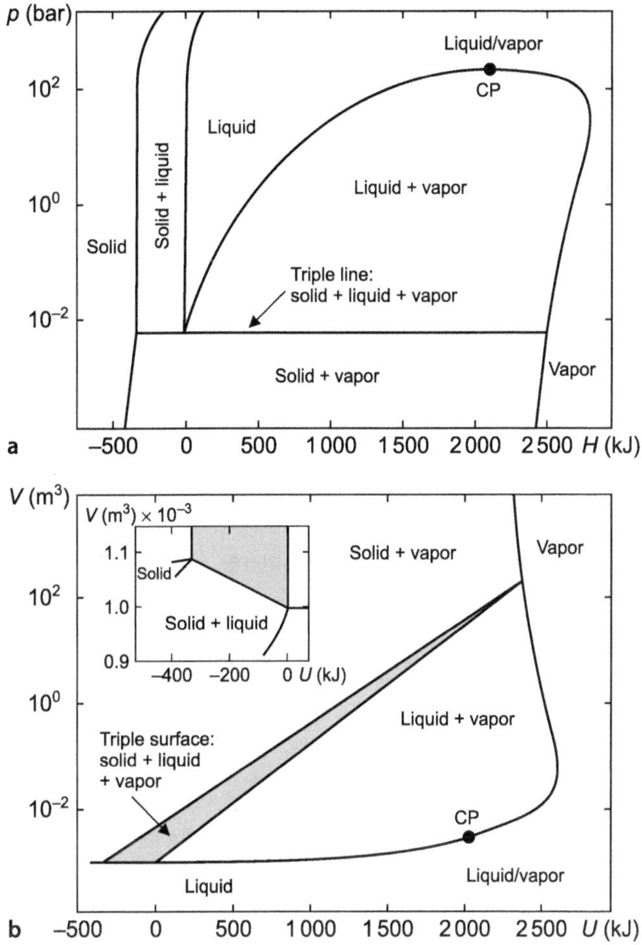

Fig. 5.6. *State diagrams of water.* Quantitative representation of single-phase, two-phase and three-phase regions in (a) the *H*-*p*-diagram and (b) the *U*-*V*-diagram. In the *H*-*p*-diagram, the triple point corresponds to a triple line. In the *U*-*V*-diagram the triple point corresponds to a triple plane. In the two-phase regions, the entropy is described by Eq. 5.41. CP is the critical point

and gives us an explicit expression for the dependence of the vapor pressure of a substance on the temperature.

Finally we would like to obtain an overview of the regions of the state space in which the description of entropy using Eq. 5.41 can be applied. In Fig. 5.6a, we plot the different states of water in an enthalpy-pressure diagram. The entropy expression (Eq. 5.41) holds for the three hatched two-phase regions. These regions limit the single-phase regions in which the substance is in the solid, liquid or gaseous state only. The two-phase regions join each other at the triple point, whose coordinates for water are $T = 0.01\,°\text{C}$ and $p = 0.00611$ bar.

Figure 5.6 shows that strictly speaking the triple point is not a point but a *triple line* in the *H*-*p*-diagram and a *triple surface* in the *U*-*V*-diagram. We have already mentioned in Sect. 2.5 (Fig. 2.4) that a unique representation of a state of a simple thermodynamic system with one work coordinate is only possible using the coordinates U and V. The example of water has shown us that this is indeed the case.

5.4
Analysis of a Vapor Power System

How can we convert the temperature difference between cold and warm ocean water, discussed in Sect. 5.1, into useful mechanical work? This can be accomplished using a power cycle. In the present section, we wish to explain how such a system works and how one can evaluate its thermodynamic performance with the help of entropy. We shall perform our analysis using a fictitious Ocean Thermal Energy Conversion (OTEC) facility. Such systems are based on the idea of the French scientist Arsene d'Arsonval. They were developed in the period between 1970 and 1985 in response to the first oil crisis (Avery and Wu 1994) and are not commercially used today (2010). We consider this academic example instead of the economically more important systems using water as the working substance. This is done in order to demonstrate that power cycles can not only work at high temperatures as they occur during the combustion of oil or coal but also with very small temperature differences. The methods of analysis, to be developed in the present section, can be applied to the analysis of coal-fired power plants or nuclear power plants with minor modifications.

A – Formulation of the Problem

Let us first translate our question into a well-defined thermodynamic problem referring to Fig. 5.7. The figure shows the principle of a special kind of power cycle, namely a *vapor power system*. A vapor power system receives energy from a reservoir with a high temperature, rejects thermal energy to a system at a low temperature and converts the difference between these energies into work. The particular property of a vapor power system is that the working fluid alternates between liquid and gaseous state. In OTEC systems, the working fluid is ammonia whereas conventional coal-fired power plants and nuclear power plants use water as the working fluid.

Warm ocean water flows through a boiler (which is also called an evaporator in OTEC systems) and transfers thermal energy to the ammonia. In order to simplify the analysis, we shall assume that the flow rate of the ocean water is so large that its temperature remains unchanged during the flow through the boiler and remains at $T_H = 20\,°C$, for example. When leaving the boiler, the internal energy of one unit of ocean water decreases by Q_H and its entropy decreases by Q_H / T_H.

We now consider the fate of a small quantity of ammonia which enters the boiler in state 1 as a liquid and leaves the boiler as a gas after having taken the thermal energy Q_H. In state 2, the ammonia has a much higher volume than previously because it is now a gas. Next, the ammonia discharges in a turbine and produces work which can be converted into electricity using a generator. The ammonia leaves the turbine with a low pressure and correspondingly low temperature. On account of the low temperature, a part of the ammonia condensates. The substance leaving the turbine is in state 3 and represents a mixture of gaseous ammonia and ammonia droplets. In the condenser the ammonia rejects the waste heat Q_C to the cold ocean water and leaves the condenser in state 4. The internal energy and entropy of the cold ocean water increase by Q_C and Q_C / T_C, respectively. Again we have assumed that the flow rate of cool ocean water is so large that the water does not change its temperature when it takes up the waste heat

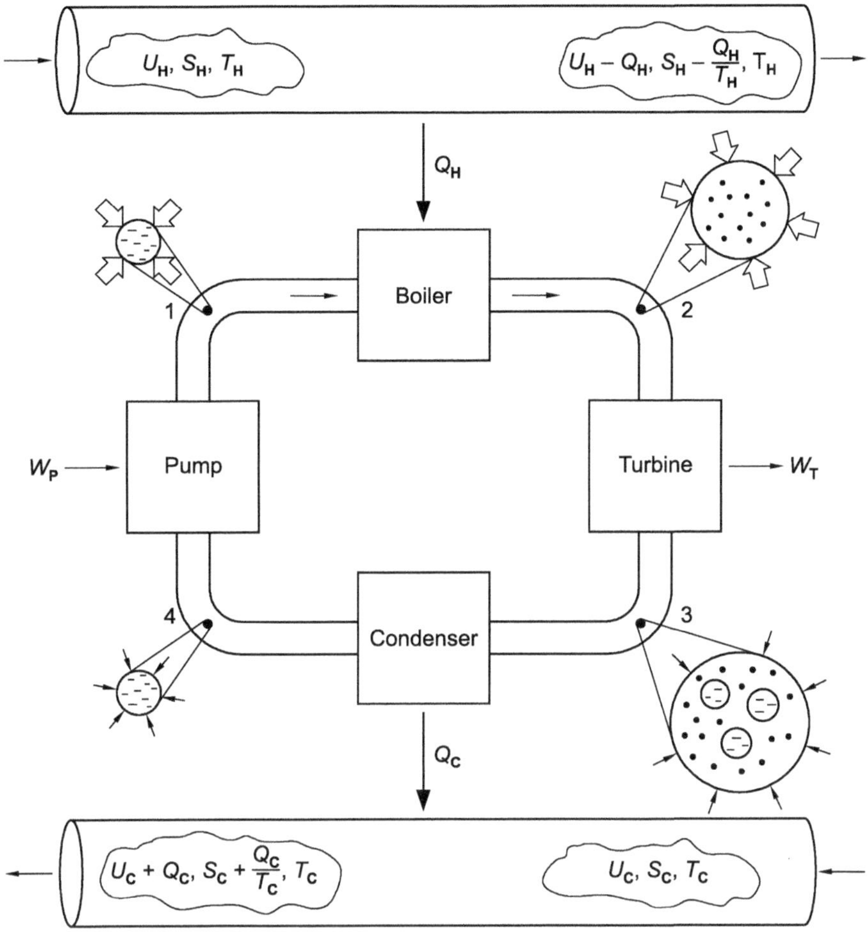

Fig. 5.7. *Principle of a vapor power system.* Sketch of the basic components of a vapor power system and of their interaction with the energy source (upper pipe) and the coolant (lower pipe). The magnified views show the states of the working fluid. The thickness of the arrows indicates the pressure. Since the volume of the working fluid remains virtually constant when it passes through the pump in the liquid state, the compression work is almost zero. Why must the pump still perform work in the amount W_P? This apparent contradiction will be resolved in Sect. 5.4 D and Fig. 5.10. The arrows marked with Q_H and Q_C indicate the exchange of energy. In reality however, the heat transfer takes place within the boiler and the condenser

from the ammonia. In the last step, a pump, which is called feedwater pump in conventional power plants, compresses the liquid ammonia to the high pressure of state 1 which is necessary to enter the boiler again. Then the ammonia repeats its cycle.

After one cycle, one element of ammonia receives the energy Q_H, rejects the energy Q_C and performs the work W. By virtue of energy conservation the work is given by $W = Q_H - Q_C$. Our goal is to compute the thermal efficiency $\eta = W/Q_H$ of the system for given values of the temperatures T_H and T_C. In order to accomplish this task, we need the thermodynamic properties of ammonia, in particular its entropy.

B – Entropy of Ammonia

In the process shown in Fig. 5.7, the ammonia is liquid in states 1 and 4, gaseous in state 2 and represents a two-phase mixture in state 3. The entropy of these states cannot be represented in one simple analytical formula as it was possible for the systems in the previous three sections. For this reason, it is common in industrial practice to use graphical representations such as the temperature-entropy diagram shown in Fig. 5.8 or tables as shown in Table 5.1. Such data are usually obtained from accurate experiments. A student who wants to work in the field of energy technology should be able to use these kinds of tables.

In the single phase regions of Fig. 5.8, marked with the labels "vapor" or "liquid," the entropy is a unique function of both temperature and pressure. For instance, the entropy of gaseous ammonia at $T = 15$ °C and $p = 6.1529$ bar is marked by the open circle at the intersection of the corresponding isobar with a fictitious horizontal line at $T = 15$ °C. The same is true for the liquid state in the left part of the diagram. For the numerical representation, it is necessary to tabulate the entropy as a function of both T and p.

In the two-phase region, which is located below the parabola-shaped line in Fig. 5.8, only one of the quantities T or p can be chosen freely. For a given temperature of $T = 20$ °C, ammonia has a vapor pressure of $p = 8.5762$ bar and the entropy is undetermined until we specify the vapor quality x. Once x is given, we can use the tabulated entropies S_f of liquid ammonia and S_g of gaseous ammonia to determine the entropy of the two-phase mixture as $S = (1 - x)S_f + xS_g$. The entropies S_f and S_g, the corresponding volumes V_f and V_g, as well as the enthalpies H_f and H_g are usually given in the form

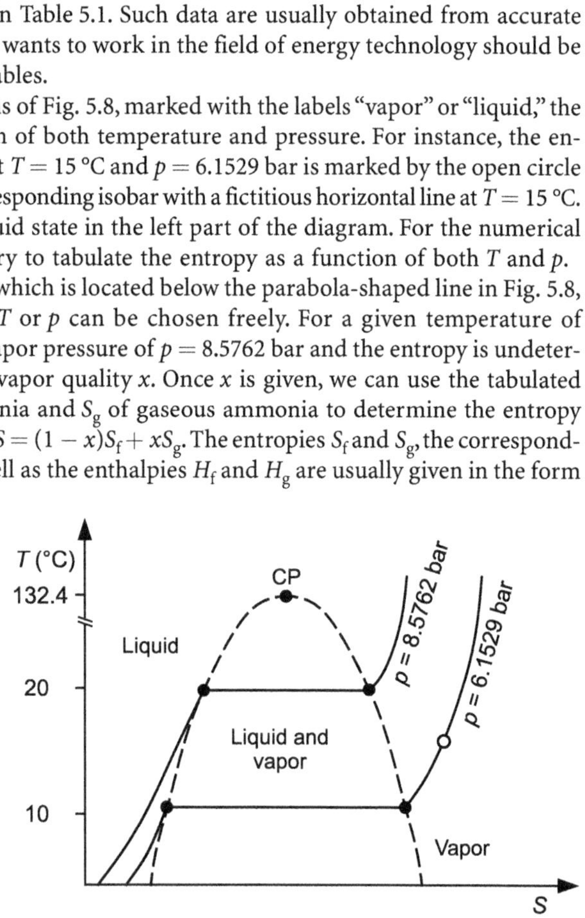

Fig. 5.8. *Entropy of ammonia.* Graphic representation of basic thermodynamic properties of ammonia. The sketch is not to scale

Table 5.1. *Entropy of ammonia.* Numerical representation of basic thermodynamic properties of ammonia. The bold numbers will be necessary for the computation of the efficiency in Sect. 5.4 C. All values have been taken from Moran and Shapiro (1995) and are given for one kilogram of substance

T (°C)	p (bar)	V_f (m^3)	V_g (m^3)	S_f (kJ K^{-1})	S_g (kJ K^{-1})	H_f (kJ)	H_g (kJ)
10	**6.1529**	**0.001601**	0.2054	**0.8769**	5.2033	**226.75**	1451.78
20	**8.5762**	0.001639	0.1492	1.0404	**5.0849**	274.26	**1459.90**

of so-called *steam tables*. Table 5.1 shows a part of a steam table which was taken from the textbook by Moran and Shapiro (1995). The numerical values are given for 1 kg of ammonia. They are given for the four points which belong to $T_C = 10$ °C and $T_H = 20$ °C and are marked with filled symbols in Fig. 5.8. The *critical temperature* $T_c = 132.4$ °C is the highest temperature at which a liquid and a gas can coexist. The critical temperature T_c (lowercase c) should not be confused with the temperature of the cold ocean water T_C (uppercase c).

In general, the computation of the thermal efficiency requires thermal property data both from the single-phase and the two-phase regions. In our analysis, we will make some simplifying assumptions which will permit us to analyze our system on the basis of two-phase data only. This makes our computation more transparent without sacrificing any important physical effect.

C – Result and Discussion

In order to compute the thermal efficiency based on the data given in Table 5.1, we need to understand where the different states are located in an *S*-*T*-diagram. We illustrate this using Fig. 5.9.

We again consider a unit of ammonia with the reference mass of 1 kg and assume that during its way through the vapor power system it is always in an equilibrium state. This is equivalent to assuming that we deal with quasistatic processes whose properties we have discussed in Sect. 4.3. Of particular use is the fact that for a quasistatic process, the quantities work and heat can be computed without any information about the properties of the surroundings of the system.

We describe the energy input to the system in the boiler as an isobaric process which transforms our ammonia from the liquid state 1 to the gaseous state 2. During the first part of this process, the temperature of the ammonia increases. This growth continues until its temperature has reached the value T_H. At this temperature, the ammonia starts to boil. Once boiling has set in, the temperature does not increase further and an increasing amount of the ammonia evaporates. This process corresponds to a horizontal movement in the *S*-*T*-diagram. We assume that at the end of the boiling process ammonia is in the form of *saturated vapor* in state 2. In general, the energy transfer from the ocean water to the ammonia is connected with entropy production because a certain temperature difference is necessary to create a heat flux. In the idealized case when this temperature difference tends to zero, the entropy production disappears as well. This case is referred to as *reversible heat transfer*. The Nobel Prize winner Richard Feynman has aptly characterized reversible heat transfer as the thermodynamic analog to the frictionless motion in classical mechanics (Feynman et al. 1963). The concept of reversible heat transfer is briefly explained in Appendix C.

Let us turn to the description of the second step which is an expansion of the ammonia vapor in the turbine. We shall assume that this is a quasistatic isentropic process. This assumption implies that we neglect the friction within the gas which is due to viscosity. After the end of the expansion process, the ammonia enters the condenser where it rejects "waste heat" to the surroundings. We assume that this is a quasistatic isobaric process. In the last step, the liquid ammonia passes a pump and its pressure increases. We shall consider this process to be isentropic and quasistatic as well.

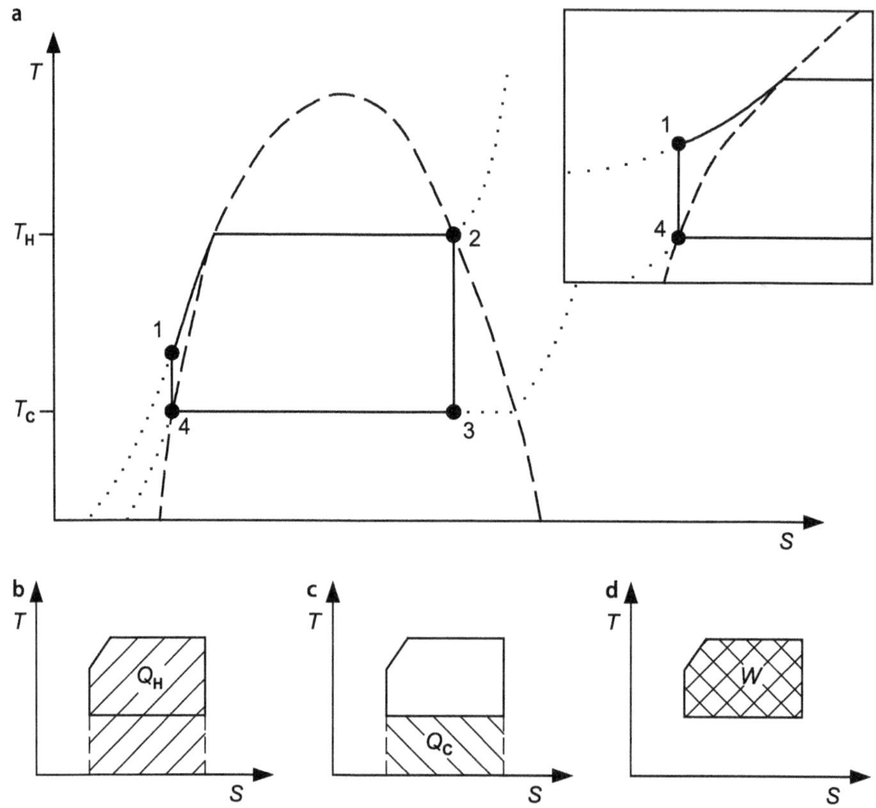

Fig. 5.9. *Rankine cycle.* (a) Representation of the path of the working substance during a Rankine cycle in an entropy-temperature diagram using the numbering introduced in Fig. 5.7. The magnifications highlight the processes taking place in the feedwater pump. The diagrams (b), (c), and (d) illustrate the geometric meanings of the energy Q_H taken from the energy source, the energy Q_C transferred to the cooling liquid, and the delivered work W respectively. The figure shows the basic variant of the Rankine process which does not involve reheat, superheat or regenerative feedwater heating

A quasistatic process for which initial and final states are identical is called a *cycle*. The present cycle is characterized by the fluid changing its state periodically between liquid and gaseous. Such a kind of cycle, which takes place in the two-phase region of a fluid and is limited by two isobars and two lines of constant entropy, is called a *Rankine cycle* and represents a very important model for industrial processes involving energy conversion.

To compute the thermal efficiency of the considered Rankine cycle, we need the quantities Q_H and Q_C, whose geometric meaning is represented in Figs. 5.9b and 5.9c, respectively. The energy taken up by the ammonia in the boiler is equal to

$$Q_H = \int_{S_1}^{S_2} TdS \qquad (5.55)$$

Using the second TdS-equation in the form $dH = TdS + Vdp$ and applying it to this isobaric process ($dp = 0$), we can readily convince ourselves that the quantity Q_H can be expressed as

$$Q_H = \int_{H_1}^{H_2} dH = H_2 - H_1 \tag{5.56}$$

Whereas H_2 is given in the Table 5.1, H_1 is located in the liquid single-phase region and is not covered by this table. However, we can use the approximation

$$H_1 \approx H_4 + (p_1 - p_4)V_4 \tag{5.57}$$

to express the enthalpy of the liquid ammonia under high pressure through its enthalpy H_4 under atmospheric pressure. Equation 5.57 can be derived from the relation $dH = Vdp$ which is the second TdS-equation in the particular case of an isentropic process. Integrating this relation and assuming that the liquid ammonia is an incompressible liquid, which obeys $V_1 = V_4$, directly yields Eq. 5.57.

The energy transferred from the ammonia to the surroundings in the condenser can be written in the form

$$Q_C = \int_{S_3}^{S_4} TdS = T_C(S_4 - S_3) \tag{5.58}$$

In order to keep the notation simple, we ignore that Q_C, as defined by this equation, has negative sign and treat it as a positive quantity which is equal to $T_C(S_3 - S_4)$. Using the last three equations, the thermal efficiency $\eta = 1 - Q_C/Q_H$ becomes

$$\eta = 1 - \frac{T_C(S_3 - S_4)}{H_2 - H_4 - (p_1 - p_4)V_4} \tag{5.59}$$

All quantities necessary to evaluate η are at our disposal and are highlighted by boldface characters in the Table 5.1. Although S_3 is not explicitly given in this table, it is readily verified that it is equal to the entropy of the saturated vapor $S_2 = S_g(20\,°\mathrm{C})$ because the expansion in the turbine is isentropic by definition. Inserting the numerical values, which correspond to 1 kg of ammonia, we obtain the expressions

$$Q_H = 1233\ \mathrm{kJ} \approx 0.34\ \mathrm{kWh} \tag{5.60}$$

$$Q_C = 1192\ \mathrm{kJ} \approx 0.33\ \mathrm{kWh} \tag{5.61}$$

$$W = 41\ \mathrm{kJ} \approx 0.011\ \mathrm{kWh} \tag{5.62}$$

which give an efficiency of

$$\eta = 3.35\% \tag{5.63}$$

This should be compared with the Carnot efficiency $\eta_{\text{Carnot}} = 1 - T_C/T_H$ whose value is 3.41%. With this result, our problem is solved. Nevertheless, two properties of the considered problem are worth being emphasized. First, the obtained thermal efficiency is very close to the Carnot efficiency. This shows that for the considered parameters the potential of the Rankine process to approach the efficiency of a Carnot cycle has been well used. This is not always the case. For Rankine cycles used in coal-fired power plants, the efficiency is much below the Carnot efficiency. Second, our example shows that in spite of sophisticated technical equipment, the overall efficiency of our process is low due to the small temperature differences.

D – Suggestions for Further Study

The work W that we have obtained as a result of our analysis is highlighted in Fig. 5.9d and can be interpreted as *net work*. However, if we try to understand this work as the difference between a *total work* delivered by the turbine and a *back work* consumed by the pump, we encounter a paradox. Since the volume of the liquid ammonia does not change during the compression between state 4 and state 1 ($V_4 = V_1$), the work performed by the pump seems to be equal to

$$W_P = -\int_{V_4}^{V_1} p\,dV = 0 \tag{5.64}$$

Why is the work of the pump equal to zero? Doesn't the pump perform the work $p_1 V_1$ in order to "push" one unit of ammonia into the boiler? And doesn't the pump receive the work $p_4 V_4$ when the ammonia is pushed into the pump by the external hydraulic circuit? And shouldn't the work of the pump therefore be equal to $W_P = p_1 V_1 - p_4 V_4 = (p_1 - p_4) V_4$?

The contradiction can be resolved by looking at Fig. 5.10, where our process is shown in a V-p-diagram. The compression work $-\int p\,dV$ performed on the liquid ammonia is indeed zero. (In reality, the work is not exactly zero due to the small but finite compressibility of the liquid. To illustrate this fact we have plotted the line which connects the points 4 and 1 in Fig. 5.10a such that it slightly deviates from the vertical.) Nevertheless, the work performed by the pump, namely $(p_1 - p_4) V_4$ is nonzero because the pump has to transport the liquid against the pressure difference $p_1 - p_4$. The quantity $(p_1 - p_4) V_4$ equals the area between the nearly vertical curve 4-1 and the V-axis. This geometrical interpretation suggests generalizing this quantity to any quasistatic process in the form

$$W_t = \int V(p)\,dp \tag{5.65}$$

In German textbooks of thermodynamics, this quantity carries the name *technische Arbeit*. This seems to be one of the few examples where a German technical term does not have an English counterpart, so we propose to use the translation *technical work* for this quantity. Figure 5.10c shows that the net work W is the difference between the magnitudes of the technical work of the steam turbine and the pump.

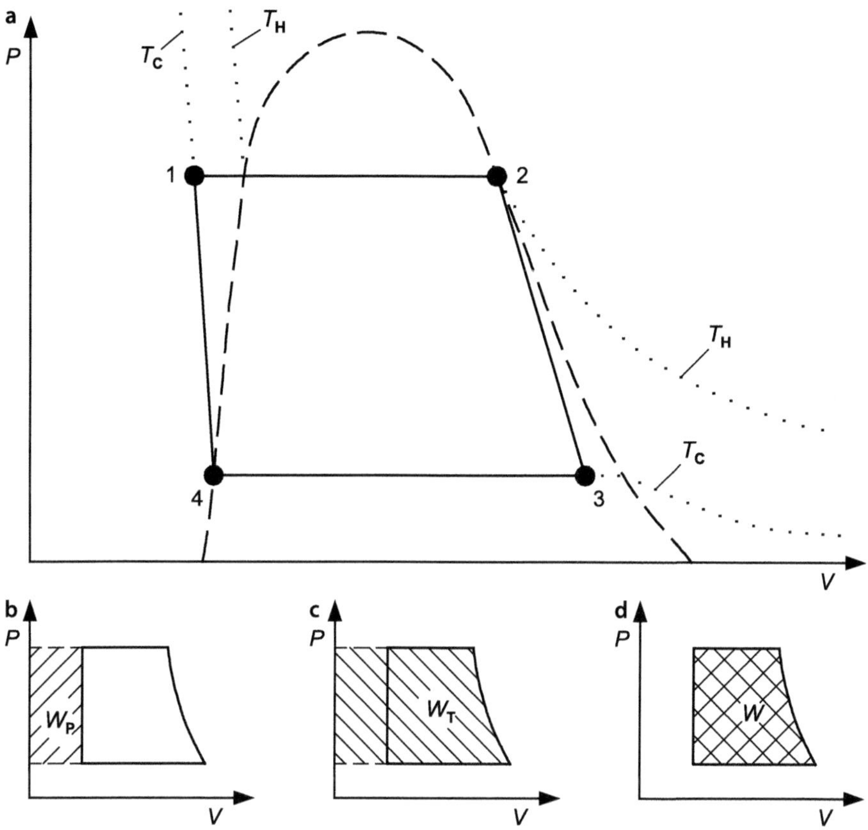

Fig. 5.10. *Technical work.* (a) Rankine process in the volume-pressure diagram. In (b) and (c) the geometric meaning of the technical work $\int V(p)\mathrm{d}p$ is illustrated as the area to the left of the $V(p)$-curve. (d) Total work as the difference between the technical work of the pump and the turbine. The line connecting points 1 and 4 would be vertical if the working fluid were incompressible when it is in a liquid state

Integrating the second $T\mathrm{d}S$-equation ($\mathrm{d}H = T\mathrm{d}S + V\mathrm{d}p$), we can furthermore convince ourselves that for a quasistatic isentropic process ($\mathrm{d}S = 0$) the technical work is equal to the difference between the enthalpies in the initial and the final states.

As a suggestion for further reading, we wish to remark that in most power plants water is used as a working fluid, not ammonia. For this reason, most thermodynamics textbooks deal with the analysis of Rankine cycles with water vapor which are used in coal-fired power plants and nuclear power plants and account for most of electrical energy that is currently produced. The temperature differences in these power plants are above 300 °C and are well suited for turbines. By contrast, for small temperature differences as in the present example, the pressure differences between the vapor pressure of the water would be as little as 0.01 bar and would be much too low for operating a turbine. Therefore, low temperature applications require working fluids like ammonia whose vapor pressure depends more strongly on temperature than for water. This example shows how important the choice of a suitable working fluid is.

5.5 Analysis of a Refrigeration System

How can we accomplish the task mentioned in Sect. 5.2 which consists of cooling hot ambient air down to temperatures suitable for air-conditioning? This is done using a system which relies on the same principles as a household refrigerator. In the present section, we wish to study the principles of operation of such systems in more detail. In particular, we will determine how much work is necessary to produce a certain quantity of "cold." As we will see, the entropy plays a central role in answering this question.

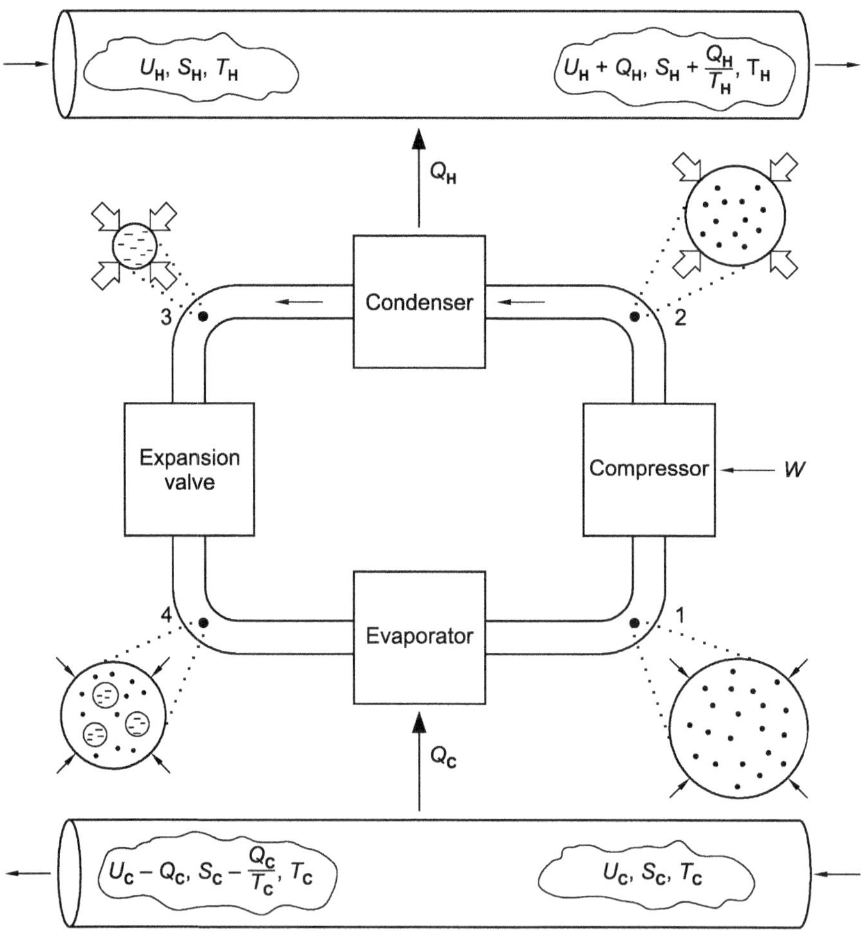

Fig. 5.11. *Principle of a refrigeration system.* Sketch of the basic components of a vapor compression refrigeration system as well as its interaction with the medium to be cooled (lower pipe) as well as the cooling fluid (upper pipe) to which the waste heat is transferred. The magnified insets show the states of the working fluid. The thickness of the arrows is proportional to the pressure. The arrows marked with Q_H and Q_C represent the energy transfer. In reality, however, the medium to be cooled is in direct contact with the evaporator and the cooling fluid flows through the condenser

A – Formulation of the Problem

We start this section by providing an accurate formulation of our problem in the language of thermodynamics. To this end, we refer to Fig. 5.11 which shows a particular kind of refrigeration system, namely a *vapor compression refrigeration system.* This technical term sounds complicated, but in fact the basic principle is well known from everyday life. When one leaves a swimming pool on a windy day, one feels cold because the water on the skin evaporates.

A vapor compression refrigeration system utilizes this effect by letting a *refrigerant* circulate in a closed circuit. Air conditioning systems of cars often use tetrafluoroethane ($C_2H_2F_4$) which is called R-134a in the terminology of refrigeration technology. R-134a is a gas which liquefies at room temperature for pressures above 6 bar.

In order to understand how the refrigeration system of Fig. 5.11 works, let us consider the fate of a small amount of refrigerant as it circulates through the system. In state 4 which represents a mixture of droplets and gas, the substance enters the *evaporator.* Here the droplets evaporate and extract the energy Q_C from the air flow which is to be cooled. Next, the pressure of the refrigerant is increased as it passes the *compressor* and is transformed into state 2. After having left the compressor, the refrigerant passes through the condenser and rejects the energy Q_H. The refrigerant leaves the condenser as a liquid. This state is marked with *3* in Fig. 5.11. Finally the refrigerant passes an *expansion valve* which often consists of a capillary and returns to its initial state 4.

Let us assume that the temperature in the evaporator is $T_C = 0$ °C and that the refrigerant leaves the condenser with a temperature $T_H = 40$ °C. In the course of a single cycle, the refrigerant has extracted the "cold" Q_C from the air to be cooled and has rejected the "waste heat" Q_H to the surroundings. The work W necessary to accomplish this task is related to these energies via the energy conservation equation $W = Q_H - Q_C$. Our goal is to compute the coefficient of performance $\beta = Q_C / W$ for given values of T_H and T_C. Before we solve this problem, we have to acquaint ourselves with the thermodynamic properties of R-134a, and in particular with its entropy.

B – Entropy of the Refrigerant R-134a

For the process shown in Fig. 5.11, the refrigerant is a gas in states 1 and 2, a liquid in state 3 and a two-phase mixture in state 4. As in Sect. 5.4, the entropy of these states cannot be expressed by a single all-encompassing analytical formula. We therefore have to turn to a graphical representation given in Fig. 5.12 and thermodynamic tables which we have summarized in Tables 5.2 and 5.3. The numerical values refer to 1 kg of refrigerant.

In contrast to the analysis in the previous section, it is not sufficient to know the thermodynamic properties for the two-phase region shown in Table 5.2. We shall also need data from the single-phase region given in Table 5.3 because the refrigerant is in the state of superheated vapor. In Table 5.3, both temperature and pressure are independent variables. As can be inferred from Table 5.2, the pressures in a liquid-gas mixture of R-134a at 0 °C and 40 °C are 2.9282 bar and 10.165 bar, respectively. The two isobars corresponding to 2.9282 bar and 10.165 bar are plotted in Fig. 5.12. For instance, if we were to determine the entropy of gaseous R-134a at 40 °C and 2.9282 bar, we would

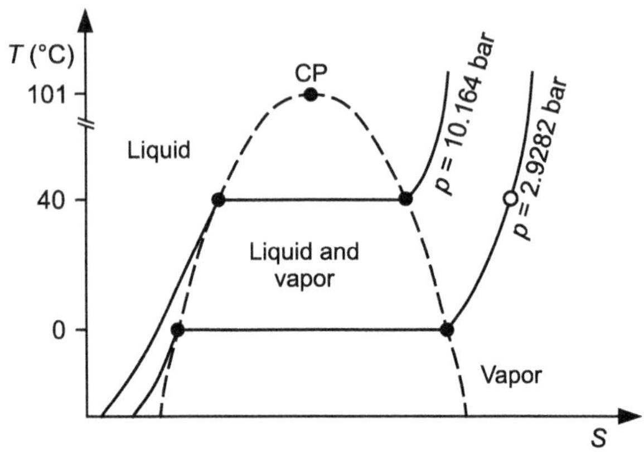

Fig. 5.12. *Entropy of the refrigerant R134a.* Graphic representation in the two-phase region. The sketch in is not to scale

Table 5.2. *Entropy of the refrigerant R134a.* Numerical representation in the two-phase region. The bold numbers are needed to calculate the coefficient of performance in Sect. 5.5 C. All numerical values have been taken from Moran and Shapiro (1995) and are given for one kilogram of R134a

T (°C)	p (bar)	V_f (m^3)	V_g (m^3)	S_f (kJ K^{-1})	S_g (kJ K^{-1})	H_f (kJ)	H_g (kJ)
0	2.9282	0.007721	0.0689	0.1970	**0.9190**	50.02	**247.2**
40	**10.164**	0.008714	0.0199	0.3866	0.9041	**106.19**	268.24

Table 5.3. *Entropy of the refrigerant R134a.* Numerical representation in the single-phase region. The bold numbers are needed to calculate the coefficient of performance in Sect. 5.5 C. All numerical values have been taken from Moran and Shapiro (1995) and are given for one kilogram of R134a

T (°C)	p (bar)	V (m^3)	S (kJ K^{-1})	H (kJ)
40	10	0.02029	**0.9066**	**268.68**
50	10	0.02171	**0.9428**	**280.19**
50	12	0.01712	**0.9164**	**275.52**

have to read the value of the abscissa which corresponds to the intersection of the 40 °C-isotherm with the 2.9282 bar-isobar which is marked with an open circle. The properties of R-134a have to be determined in a similar way to those of ammonia in the previous section.

C – Result and Discussion

To compute the coefficient of performance of our refrigeration system based on the thermodynamic data of Fig. 5.12 and Tables 5.2 and 5.3, we have to locate the states shown in Fig. 5.11 in the *S*-*T*-diagram. This task is accomplished with the help of Fig. 5.13.

As in the previous section, we shall assume that all processes are quasistatic. We shall further assume that the compression $1 \rightarrow 2$ is an isentropic process which starts at the 2.9282 bar-isobar and ends at the 10.164 bar-isobar. The compressor must hence

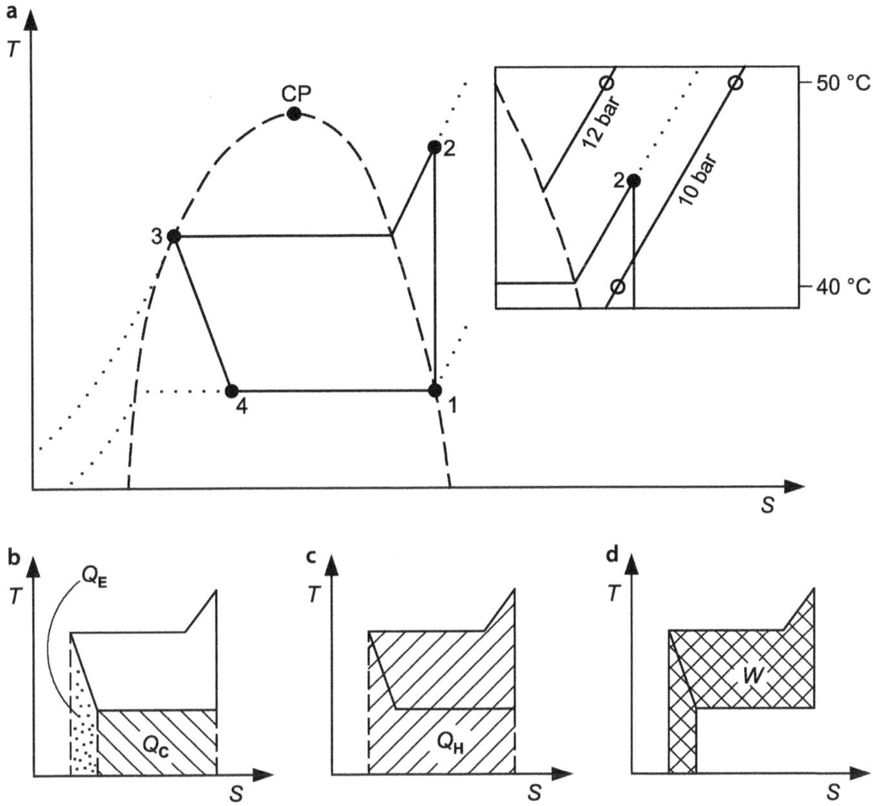

Fig. 5.13. *Vapor compression refrigeration system.* (a) Representation of the path of the working substance in the entropy-temperature diagram using the numbering introduced in Fig. 5.11. The inset shows the numerical values as open circles from which we interpolate the enthalpy in point 2. The diagrams (b), (c), and (d) illustrate the geometric meaning of the energy Q_C taken from the cold reservoir, the energy Q_H discharged into the surroundings and the work W, respectively. The energy Q_E shown in (b) is generated in the expansion valve and is equal to the technical work for the transition from state 3 to state 4 (cf. Sect. 5.4 C, in particular Fig. 5.10 and Eq. 5.65). CP denotes the critical point

overcome a pressure difference of 7.2358 bar. The refrigerant leaves the compressor as superheated vapor whose temperature is assumed to be higher than 40 °C. We also assume that the refrigerant remains at constant pressure in the condenser. The energy transfer is completed once the refrigerant is entirely in the liquid state. This is the case in state 3. Next we let the refrigerant pass through the expansion valve without exchanging energy with the surroundings. During this process, the temperature is assumed to fall to 0 °C, corresponding to state 4. Now the refrigeration process itself starts: The refrigerant boils, takes up energy and thereby produces the desired "cold." Finally, the refrigerant leaves the evaporator in state 1.

As in the case of vapor power systems considered in Sect. 5.4, both the heat transfer to the refrigerant and from the refrigerant occur at constant pressure. Hence we can express the energies in an analogous way as

$$Q_H = H_2 - H_3 \tag{5.66}$$

$$Q_C = H_1 - H_4 \tag{5.67}$$

Whereas the states 1, 2 and 3 are uniquely determined, one piece of information necessary to locate state 4 is missing. In order to obtain the missing piece, we make use of the assumption that the refrigerant does not exchange energy with its surroundings when passing the expansion valve. The internal energy of a given amount of the refrigerant changes by $p_3 V_3$ because it is pushed into the valve and it changes by $-p_4 V_4$ because is pushes other fluid particles when leaving the valve. Hence $U_4 - U_3 = p_3 V_3 - p_4 V_4$ which is equivalent to $U_4 + p_4 V_4 = U_3 + p_3 V_3$ and, by virtue of the definition of enthalpy $H = U + pV$, to

$$H_3 = H_4 \tag{5.68}$$

We have thus shown that the expansion in the valve is an *isenthalpic process*. We can therefore write the coefficient of performance $\beta = Q_C / W = Q_C / (Q_H - Q_C)$ as

$$\beta = \frac{H_1 - H_3}{H_2 - H_1} \tag{5.69}$$

Our remaining task is to compute the enthalpies H_1, H_2 and H_3. This computation does not require any new physics but a little bit of elementary mathematics. We assume that our refrigeration system contains 1 kg of refrigerant. From Table 5.2, we immediately obtain $H_1 = 247.2$ kJ. The determination of H_2 requires somewhat more effort, since we do not know the temperature T_2 at the end of the compression process. Here entropy is of great help, as the inset in Fig. 5.13 shows. Since the transition $1 \rightarrow 2$ is an isentropic process, $S_2 = S_1 = 0.919$ kJ K^{-1}. From Table 5.3 we know the enthalpy for three pairs (S, p) which are plotted as three open circles in the inset of Fig. 5.13a. These numerical values are sufficient to determine the value of H for $S = 0.919$ kJ K^{-1} and $p = 10.164$ bar through interpolation.

To carry out the interpolation, we need to perform the following auxilary computation: We approximate the unknown function $H(S, p)$ by a plane which is defined by the three values of S, p and H given in the table. The general equation of the plane is

$$H(S, p) = a + bS + cp \tag{5.70}$$

To determine the three unknown coefficients a, b, and c, we insert the given numerical values and obtain the linear system of equations

$$268.68 = a + 0.9066b + 10c \tag{5.71}$$

$$280.19 = a + 0.9428b + 10c \tag{5.72}$$

$$275.52 = a + 0.9164b + 12c \tag{5.73}$$

Since the determination of a, b and c is a purely mathematical task, we ignore all physical units for a moment and treat the numerical values as non-dimensional quantities. Equations 5.71–5.73 form a linear algebraic system for a, b and c, whose solution we leave to the reader as an exercise. The result $a \approx 38.204$, $b \approx 317.96$, $c \approx 1.8621$ permits us to use Eq. 5.70 to determine the desired value of the enthalpy for $S = 0.919$ kJ K^{-1} and $p = 10.164$ bar as $H_2 = 272.9$ kJ. As a by-product, we have performed a similar interpolation and determined the temperature at point 2 as $T_2 = 44$ °C. This quantity, however, is only mentioned for illustrative purposes. It is not needed for the computation of β.

The last missing value of the enthalpy, namely for point 3, is quickly obtained from Table 5.2 as $H_3 = 106.19$ kJ. Using Eq. 5.66 we obtain $Q_C = 141$ kJ, $Q_H = 167$ kJ, $W = 26$ kJ and the desired coefficient of performance becomes

$$\beta = 5.42 \tag{5.74}$$

This is our desired result. For comparison, let us compute the coefficient of performance of a Carnot refrigeration cycle:

$$\beta_{\text{Carnot}} = \frac{T_C}{T_2 - T_C} = 6.21 \tag{5.75}$$

Observe that we have to use $T_2 - T_C$ in computing the Carnot coefficient of performance, not $T_H - T_C$, because T_2 is the maximum temperature in our process, as seen in Fig. 5.13a. As expected, the coefficient of performance for the Carnot refrigeration cycle is higher than for the real vapor compression refrigeration cycle. The coefficient of performance shows us that for the given parameters one kilowatt of compressor power can produce a *refrigeration capacity* of 5.42 kW. With this result our problem is solved.

D – Suggestions for Further Study

It is instructive to consider the isenthalpic process 3 $\rightarrow$ 4, which takes place in the expansion valve, in some more detail. In contrast to the turbine and pump discussed in Sect. 5.4 D, the technical work in the expansion valve $W_t = \int V(p)\mathrm{d}p$ (see Sect. 5.4 D for the introduction of the term *technical work*) is not transferred to the surroundings. It is immediately transformed into "heat" which stays inside the system and is denoted as Q_E in Fig. 5.13b where the index E stands for expansion. An integration of the second $T\mathrm{d}S$-equation shows that $|Q_E| = |W_t|$. This implies that "valuable" work is transformed into "useless" heat. This fact has adverse consequences for the coefficient of performance, as illustrated in Fig. 5.13d. Indeed, this figure shows that it is necessary to perform more work than for a system in which the process 3 $\rightarrow$ 4 is isentropic.

Up to now have we have not asked the question how the temperatures T_C and T_H have to be chosen in practice. The method for choosing these quantities in the course of the design of a refrigeration system provide an illustrative example of how closely engineering thermodynamics is related to the subject of heat transfer. If the maximization of the coefficient of performance were the only design objective, one would have to use the lowest possible condensation temperature T_H and the highest possible evaporation temperature T_C. This would be the case if T_C were identical with the desired low temperature, for instance the desired temperature of a bottle of beer in the refrigerator, and if T_H were equal to the temperature of the surroundings, for instance the room temperature for a household refrigerator. However, as explained in Appendix C, the transfer of the energies Q_H and Q_C between our system and its surroundings would take infinitely long. According to the science of heat transfer, which (in contrast to thermodynamics) does deal with the speed of transfer processes, it is advantageous if the temperature differences between the condenser and the surroundings on the one hand and between the evaporator and the cooling chamber on the other hand, are as high as possible. This requirement in turn reduces the coefficient of performance. The contradicting requirements of a high coefficient of performance and high heat transfer rates are treated in detail in textbooks on refrigeration technology which are recommended to the reader for further study.

Finally it should be noted that apart from vapor compression refrigeration systems, there are at least two other important techniques for refrigeration, namely *absorption refrigeration systems* and *gas refrigeration systems*. Absorption refrigeration systems are capable of producing low temperatures with a small amount of mechanical work. They are therefore particularly suited for the use of waste heat in power plants or for the use of solar energy. In absorption refrigeration systems, a mixture of water and ammonia is often used as a working fluid. Gas refrigeration systems employ substances like helium which stay in the gaseous state and are therefore suitable for generating the very low temperatures that are necessary for the cooling of superconducting magnets.

5.6 Production of Ammonia

People in developed countries enjoy a plentiful supply of food. However, as recent as in the twentieth century, and all the more in the nineteenth century, people were often plagued with famine, which was primarily caused by the lack of scientific methods for the increase of agricultural productivity. It is thanks to ammonia as a basis for the production of synthetic fertilizers that modern agriculture has relieved us from the troubles our ancestors had with starvation. Ammonia is one of the most important raw materials in the chemical industry. Most ammonia is produced from nitrogen and hydrogen using the Haber-Bosch process. The method is embodied in the chemical reaction

$$N_2 + 3H_2 \rightleftharpoons 2NH_3 \tag{5.76}$$

We are interested in the question about the optimal values of the temperature and pressure. We would also like to know the maximum amount of ammonia which can be produced from a given quantity of nitrogen and hydrogen. These questions will be addressed in the present section.

A – Formulation of the Problem

Let us first translate the foregoing questions into a well-defined thermodynamic problem. Referring to Fig. 5.14a, we consider a vessel which is filled with one mole of nitrogen and three moles of hydrogen whose temperature and pressure are given. If the system

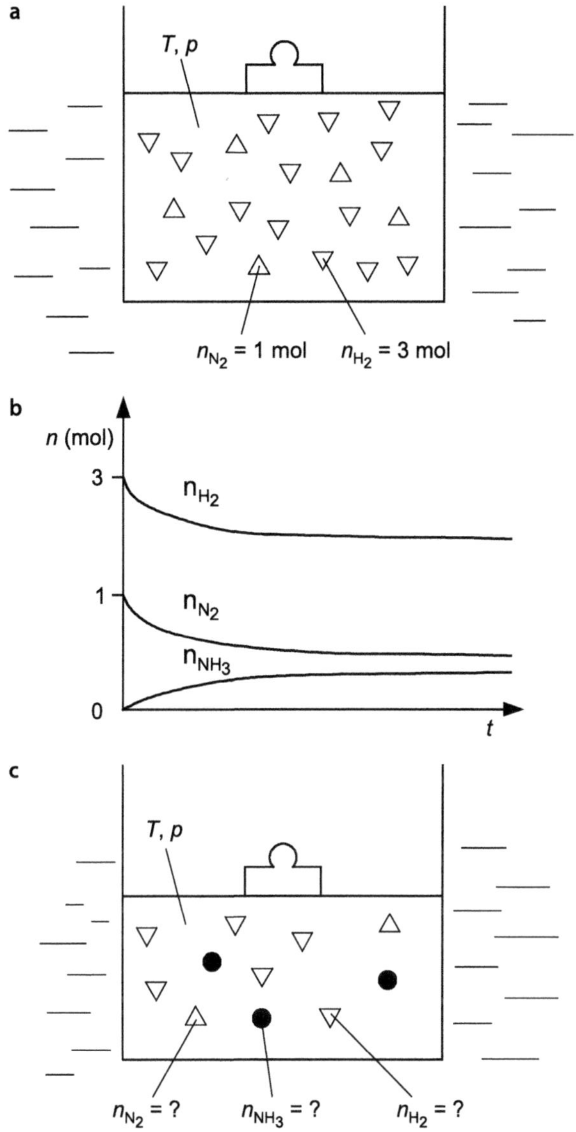

Fig. 5.14.
Synthesis of ammonia. (a) Initial state; (b) mole numbers during the reaction as a function of time; (c) equilibrium state. Our goal is to compute the mole numbers n_{N_2}, n_{H_2}, and n_{NH_3} in the thermodynamic equilibrium which is established after a sufficiently long time when the values of T and p are held constant. The initial state is not an equilibrium state. Therefore, strictly speaking, no entropy can be associated with this state. Nevertheless, it is instructive for the purpose of comparison to compute its hypothetic entropy if it were an equilibrium state

consists of nitrogen and hydrogen only, it is not in a thermodynamic equilibrium state. Even if pressure and temperature are held constant, the system does not remain in this state. Indeed, the two gases react and form ammonia. If we could measure the quantities of the individual gases as functions of time, we would obtain a result similar to that depicted in Fig. 5.14b. After a certain time, whose duration cannot be predicted by thermodynamics, the system would settle at the equilibrium state sketched in Fig. 5.14c. In this state, the quantities of the gases are constant and do not change with time. We wish to determine the number of moles n_{N_2}, n_{H_2}, and n_{NH_3} in this equilibrium state.

At first glance, the problem can be solved by writing down the entropy $S(T, p, n_{N_2}, n_{H_2}, n_{NH_3})$ of the state shown in Fig. 5.14c and searching for the maximum of S while keeping T, p as well as the number of nitrogen atoms $N_{N_2} = 2 \cdot n_{N_2} + 1 \cdot n_{NH_3}$ and hydrogen atoms $N_{H_2} = 2 \cdot n_{H_2} + 3 \cdot n_{NH_3}$ constant. This procedure, however, does not lead to the correct result because the transition from the initial state shown in Fig. 5.14a to the equilibrium state shown in Fig. 5.14c is not an adiabatic process. We can therefore not directly apply the law of entropy increase derived in Sect. 3.4. The reaction is *exothermic*, i.e., the system rejects heat to the environment in order to keep the temperature constant.

In order to be able to apply the law of entropy increase to determine the equilibrium state, we have to include both the vessel and its surroundings into our analysis. This is done in Fig. 5.15. We assume that the temperature in the vessel is kept constant by connecting it to a reservoir which is so large that its temperature remains unchanged no matter what takes place inside the vessel. We further assume that the compound

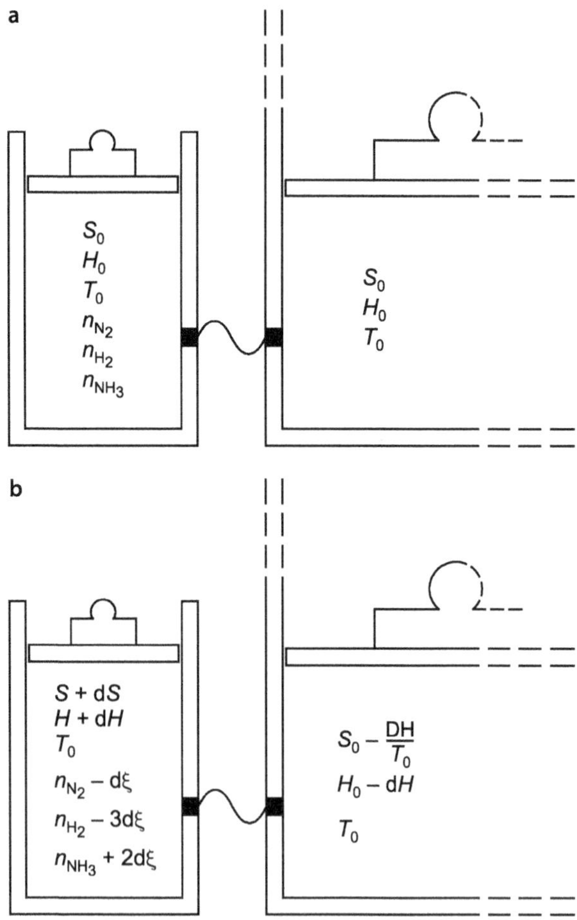

Fig. 5.15. *Minimum principle for the Gibbs function.* A mixture of nitrogen, hydrogen, and ammonia undergoes an infinitesimal adiabatic process from state (a) to state (b). The temperature and pressure are held constant. The entropy of the compound system (a) consisting of the gas mixture and the reservoir is maximum if the Gibbs function of the gas mixture attains a minimum with respect to the extent of the reaction ξ. This statement holds independently of the particular properties of the reservoir. It also holds if the reservoir consists of a substance that is held at constant volume rather than at constant pressure. Notice that only one of the states (a) or (b) can be an equilibrium state with a well-defined entropy. Nevertheless, it makes sense to assume that both states have an entropy in order to be able to compare them

system consisting of the vessel and the reservoir perform a process similar to a transition between the states shown in Fig. 5.15a and b but with the particular property that the initial and the final states differ only by an infinitesimal amount of the entropy. We can then apply the law of entropy increase to this infinitesimal process and answer the question whether redistribution between the gases N_2 and H_2 on the one hand and the reaction product NH_3 on the other hand increases the entropy of the compound system.

During the process that takes the system from the state shown in Fig. 5.15a to the state in Fig. 5.15b, the number of moles of N_2 decreases by $d\xi$. The parameter ξ is sometimes referred to as the *extent of reaction*. Since one molecule of nitrogen reacts with three molecules of hydrogen, the amount of hydrogen decreases by $3d\xi$ whereas the amount of ammonia increases by $d\xi$. Furthermore, the enthalpy of the compound system remains unchanged because the pressure is constant and exchange of energy occurs only between the two subsystems, not between the system and its surroundings. It follows that an increase of the enthalpy in the vessel by dH is accompanied by a change of enthalpy in the reservoir by $-dH$.

Based on these considerations, we can compute the change of the entropy of the compound system. The entropy of the compound system in the initial state is

$$S + S_0 \tag{5.77}$$

where S is the entropy of the considered system and S_0 the entropy of the reservoir. In the final state, the entropy is equal to

$$S + dS + S_0 - \frac{dH}{T} \tag{5.78}$$

In writing down this equation, we have used the fact derived in Sect. 5.2 that the entropy of a system with constant temperature and constant pressure changes by dH/T if its enthalpy increases by dH. By comparing Eqs. 5.77 and 5.78, we conclude that the entropy experiences a change of

$$\Delta S = dS - \frac{dH}{T} \tag{5.79}$$

According to the law of entropy increase, the entropy of a thermodynamic system cannot decrease in an adiabatic process which is expressed by $\Delta S \geq 0$. Hence the considered infinitesimal process must obey

$$dS - \frac{dH}{T} \geq 0 \tag{5.80}$$

This is equivalent to

$$dH - TdS \leq 0 \tag{5.81}$$

On the left-hand side of this inequality, we recognize the differential of the Gibbs function $G = H - TS$ (cf. Sect. 4.6 D) evaluated at constant temperature. We are thus led to the important result

$$dG \leq 0 \tag{5.82}$$

which is a consequence of the entropy principle. This inequality tells us that the Gibbs function cannot increase in the considered process. If the initial state shown in Fig. 5.15a is already a minimum of G with respect to the extent of reaction ξ, the system cannot spontaneously leave this state. Consequently, it must be the equilibrium state in question. We can summarize this result as follows.

Minimum principle of the Gibbs function. For given values of temperature and pressure, a thermodynamic system is in thermodynamic equilibrium if the Gibbs function has a minimum with respect to the remaining coordinates.

A word of caution is in order regarding the interpretation of this minimum principle. One is often tempted to ascribe a mysterious "desire" to thermodynamic systems to evolve into a state with minimum Gibbs function similar to a sleigh which moves downhill in order to minimize its potential energy. In contrast to a sleigh, whose state is governed by the laws of classical mechanics no matter whether it moves or stands still, classical thermodynamics applies exclusively to equilibrium states comparable to the stable equilibrium of a sleigh in a valley or its unstable equilibrium at the top of a hill. Classical thermodynamics cannot therefore make predictions about the speed of processes, not to mention mysterious desires. The desire of a vessel containing a mixture of oxygen and hydrogen for the redemptive chemical reaction into water will remain unfulfilled even if an observer is willing to wait one million years. The fact that a burning match can initiate this chemical reaction is strictly speaking not the subject of thermodynamics. The minimum principle merely asserts that water cannot spontaneously transform into a mixture of oxygen and hydrogen as a result of an adiabatic process.

In order to predict the equilibrium state in Fig. 5.14c, we thus have to determine the Gibbs function $G(T, p, n_{N_2}, n_{H_2}, n_{NH_3})$ of our system, to express the numbers of moles through the extent of reaction ξ and to find the minimum of G with respect to this parameter. In order to accomplish this task, it is necessary to perform an intermediate step which consists in the computation of entropy and of the Gibbs function for a mixture of several gases.

B – Entropy and Gibbs Function of a Mixture of Ideal Gases

The mixture shown in Fig. 5.14 consists of three gases. When developing a theory, it is desirable to start with the simplest possible case. We therefore assume that the gases in our example behave according to the ideal gas law. Let us first derive the entropy and the Gibbs function for a mixture of two ideal gases denoted by A and B. The generalization to three or more gases will turn out to be straightforward. We already know the entropy of an ideal gas from our analysis in Sect. 5.2 B (Eq. 5.24). We rewrite this equation in the form

$$S(T,p) = n\left[s_0 + c_p \ln(T/T_0) - \mathrm{R}\ln(p/p_0)\right] \tag{5.83}$$

Remember that s_0, T_0 and p_0 are constants which can be chosen arbitrarily, for instance as $s_0 = 0$, $T_0 = 273.15$ K and $p_0 = 1$ bar.

Since A and B are ideal gases, each of them behaves as if it were alone within the volume V. Hence the individual gases obey the thermal equations of state $p_A V = n_A \mathrm{R}T$ and $p_B V = n_B \mathrm{R}T$ where the pressures p_A and p_B are called *partial pressures* of the components A and B. The total pressure $p = p_A + p_B$ is a sum of the partial pressures. By adding the two thermal equations of state we conclude that the thermal equation of state of the mixture is $pV = n\mathrm{R}T$ where $n = n_A + n_B$ is the total number of moles. Moreover, by dividing each thermal equation of state by the thermal equation of the mixture, we obtain $p_A = (n_A/n)p$ and $p_B = (n_B/n)p$. Since the gases do not interact with each other, their entropies are

$$S_A = n_A\left[s_{0A} + c_{pA} \ln(T/T_0) - \mathrm{R}\ln(p_A/p_0)\right] \tag{5.84}$$

$$S_B = n_B\left[s_{0B} + c_{pB} \ln(T/T_0) - \mathrm{R}\ln(p_B/p_0)\right] \tag{5.85}$$

and the total entropy is given by the sum $S = S_A + S_B$ of the entropies of the individual gases. The total entropy $S = S_A + S_B$ is therefore

$$\begin{aligned} S(T,p,n_A,n_B) = {} & n_A\left[s_{0A} + c_{pA} \ln(T/T_0) - \mathrm{R}\ln(p/p_0)\right] \\ & + n_B\left[s_{0B} + c_{pB} \ln(T/T_0) - \mathrm{R}\ln(p/p_0)\right] \\ & - \mathrm{R}\left[n_A \ln(n_A/n) + n_B \ln(n_B/n)\right] \end{aligned} \tag{5.86}$$

The structure of this formula becomes more transparent if we use the symbolic notation

$$S = n_A s_A + n_B s_B + \Sigma \tag{5.87}$$

This result can be summarized as follows.

Entropy of a mixture of ideal gases. The entropy of a mixture of ideal gases is given by the sum of their individual entropies plus a quantity Σ which is called the mixing entropy. The mixing entropy depends only on the numbers of moles of the gases and for the particular case of two gases takes the form $\Sigma(n_A, n_B) = -\mathrm{R}[n_A \ln(n_A/n) + n_B \ln(n_B/n)]$ with $n = n_A + n_B$.

The mixing entropy is always positive. This indicates that the mixture of two ideal gases is always an adiabatic process. The isolines of the mixing entropy are shown in Fig. 5.16 for the case of two gases. The most important properties of the mixing entropy are summarized in Appendix D. This appendix also contains an explanation why it is appropriate to call this quantity *mixing* entropy.

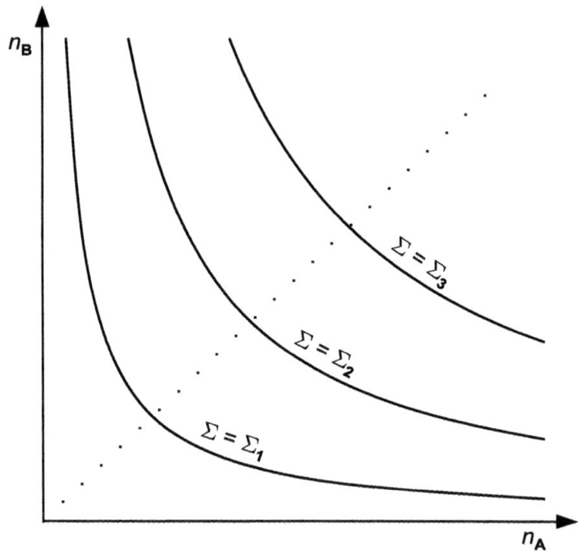

Fig. 5.16.
Mixing entropy of two ideal gases. The mixing entropy $\Sigma(n_A, n_B)$ can be visualized as a hill whose lines of constant height $\Sigma_1 < \Sigma_2 < \Sigma_3$ are convex, grow in "northeastern" direction and which is symmetric with respect to the line $n_A = n_B$

To complete the computation of the Gibbs function $G = H - TS$ of the considered gas mixture we need its enthalpy. The enthalpy of an ideal gas whose molar heat capacity c_p is independent of the temperature, reads

$$H(T) = n\left[h_0 + c_p(T - T_0)\right] \tag{5.88}$$

Hence the enthalpy of the mixture is given by the sum

$$H(T) = n_A\left[h_{0A} + c_{pA}(T - T_0)\right] + n_B\left[h_{0B} + c_{pB}(T - T_0)\right] \tag{5.89}$$

and the Gibbs function becomes

$$\begin{aligned} G(T,p,n_A,n_B) = {} & n_A\left\{h_{0A} + c_{pA}(T - T_0) - T\left[s_{0A} + c_{pA}\ln(T/T_0) - R\ln(p/p_0)\right]\right\} \\ & + n_B\left\{h_{0B} + c_{pB}(T - T_0) - T\left[s_{0B} + c_{pB}\ln(T/T_0) - R\ln(p/p_0)\right]\right\} \\ & + RT\left[n_A\ln(n_A/n) + n_B\ln(n_B/n)\right] \end{aligned} \tag{5.90}$$

This expression can be simplified using the symbolic notation

$$G = n_A g_A + n_B g_B + T\Sigma \tag{5.91}$$

and represents the central result of this subsection. It can be summarized as follows.

Gibbs function of a mixture of ideal gases. The Gibbs function of a mixture of ideal gases is equal to the sum of the Gibbs functions of its constituents minus the term $T\Sigma$ where Σ represents the mixing entropy.

C – Result and Discussion

After having derived the Gibbs function for a mixture of two arbitrary ideal gases, let us return to our original problem of determining the numbers of moles n_{N_2}, n_{H_2} and n_{NH_3} in the chemical equilibrium during the production of ammonia. This requires a generalization of our expression to the case of three ideal gases. In order to avoid cumbersome double indices, we shall from now on replace the notation n_{N_2}, n_{H_2} and n_{NH_3} with n_A, n_B and n_C and use the symbols A, B and C as representatives for nitrogen, hydrogen and ammonia, respectively. Starting from Eq. 5.91 we can readily generalize the Gibbs function to three gases in the form

$$G = n_A g_A + n_B g_B + n_C g_C + T\Sigma \tag{5.92}$$

where the Gibbs functions of the individual gases are

$$\begin{aligned} g_A(T,p) &= h_{0A} + c_{pA}(T - T_0) - T\left[s_{0A} + c_{pA}\ln(T/T_0) - \mathrm{R}\ln(p/p_0)\right] \\ g_B(T,p) &= h_{0B} + c_{pB}(T - T_0) - T\left[s_{0B} + c_{pB}\ln(T/T_0) - \mathrm{R}\ln(p/p_0)\right] \\ g_C(T,p) &= h_{0C} + c_{pC}(T - T_0) - T\left[s_{0C} + c_{pC}\ln(T/T_0) - \mathrm{R}\ln(p/p_0)\right] \end{aligned} \tag{5.93}$$

and the mixing entropy is

$$\Sigma(n_A, n_B, n_C) = -\mathrm{R}\left[n_A \ln(n_A/n) + n_B \ln(n_B/n) + n_C \ln(n_C/n)\right] \tag{5.94}$$

where $n = n_A + n_B + n_C$.

When searching for the minimum of G, we have to take into account that the numbers of moles of nitrogen, hydrogen and ammonia cannot be varied independently. For instance, if the amount of nitrogen decreases by 1 mol, the reaction Eq. 5.76 indicates that the quantity of hydrogen decreases by 3 mol and the amount of ammonia increases by 2 mol. We incorporate this fact into our analysis by using the previously defined extent of reaction ξ. Our Reaction 5.76 represents a particular case of the general reaction equation

$$a\mathrm{A} + b\mathrm{B} \rightleftharpoons c\mathrm{C} \tag{5.95}$$

with $a = 1$, $b = 3$ and $c = 2$. The extent of reaction ξ, also called reaction parameter, describes all possible combinations of numbers of moles. This includes the case $n_A = a$, $n_B = b$, $n_C = 0$ where $\xi = 0$ and we have raw materials only. It also includes the case $n_A = 0$, $n_B = 0$, $n_C = c$ where $\xi = 1$ and we have the reaction product only. In case of a *stochiometric mixture* the numbers of moles of nitrogen and hydrogen are present in the ratio a/b such that they can completely react to form ammonia. Then the numbers of moles are given by

$$n_A = a(1 - \xi), \quad n_B = b(1 - \xi), \quad n_C = c\xi \tag{5.96}$$

In an infinitesimal step of reaction they change by

$$\mathrm{d}n_\mathrm{A} = -a\mathrm{d}\xi, \quad \mathrm{d}n_\mathrm{B} = -b\mathrm{d}\xi, \quad n_\mathrm{C} = c\mathrm{d}\xi \tag{5.97}$$

The task of determining the equilibrium state consists now in finding the Gibbs function as a function of the extent of reaction in the form $G(T, p, \xi)$ and to search for its minimum with respect to ξ while keeping T and p constant. This requires finding the zero of the derivative $\partial G/\partial\xi$. Carrying out the differentiation using the chain rule we obtain

$$\frac{\partial G}{\partial \xi} = \left(\frac{\partial G}{\partial n_\mathrm{A}}\right)\frac{\mathrm{d}n_\mathrm{A}}{\mathrm{d}\xi} + \left(\frac{\partial G}{\partial n_\mathrm{B}}\right)\frac{\mathrm{d}n_\mathrm{B}}{\mathrm{d}\xi} + \left(\frac{\partial G}{\partial n_\mathrm{C}}\right)\frac{\mathrm{d}n_\mathrm{C}}{\mathrm{d}\xi} = 0 \tag{5.98}$$

After a lengthy but straightforward computation using the Relations 5.92, 5.93, 5.94 and 5.97 we are led to

$$-a\left[g_\mathrm{A} + RT\ln(n_\mathrm{A}/n)\right] - b\left[g_\mathrm{B} + RT\ln(n_\mathrm{B}/n)\right] + c\left[g_\mathrm{C} + RT\ln(n_\mathrm{C}/n)\right] = 0 \tag{5.99}$$

which can be rewritten as

$$\frac{\left(\dfrac{n_\mathrm{C}}{n}\right)^c}{\left(\dfrac{n_\mathrm{A}}{n}\right)^a\left(\dfrac{n_\mathrm{B}}{n}\right)^b} = K(T,p) \tag{5.100}$$

with

$$K(T,p) = \exp\left(\frac{ag_\mathrm{A} + bg_\mathrm{B} - cg_\mathrm{C}}{RT}\right) \tag{5.101}$$

Equation 5.100 is the main result of our analysis. It is referred to as the *law of mass action*. The validity of this law is by no means restricted to stochiometric mixtures. It states that the concentration of the reactants and products of reaction are functions of temperature and pressure only. The function $K(T, p)$ appearing on the right-hand side of Eq. 5.100 is called the *equilibrium constant*. If the equilibrium constant increases due to changes of temperature or pressure, the concentration of the reaction products appearing in the numerator of Eq. 5.100 must increase as well. Conversely, if K decreases, the thermodynamic equilibrium is shifted such that the concentration of the reactants appearing in the denominator of Eq. 5.100 increases.

If the Gibbs functions of all substances participating in the chemical reaction are known, one can determine the equilibrium constant which in turn provides the concentration of all gases in the following way.

For the production of ammonia we have $a = 1$, $b = 3$ and $c = 2$. Hence Eq. 5.96 takes the form

$$n_A = 1 - \xi, \quad n_B = 3 - 3\xi, \quad n_C = 2\xi \tag{5.102}$$

The total number of moles $n = n_A + n_B + n_C$ can be expressed as $n = 4 - 2\xi$ where ξ represents the previously defined extent of reaction. Inserting these relations into the left-hand side of Eq. 5.100, we obtain the relation

$$\frac{16\xi^2(2-\xi)^2}{27(1-\xi)^4} = K(T,p) \tag{5.103}$$

This is an implicit nonlinear algebraic equation whose solution, for given values of temperature and pressure, provides the extent of reaction $\xi(T,p)$ in thermodynamic equilibrium. Once this quantity is known, Eq. 5.102 can be used to compute the numbers of moles of hydrogen, nitrogen and ammonia in the equilibrium state. Table 5.4 shows the result which would have been obtained if we had actually carried out such a computation. With this result our task is fulfilled. It is remarkable that the content of such an extensive table can be described by a comparatively simple formula.

Table 5.4 reflects an important property of Reaction 5.76. The concentration of ammonia increases with increasing pressure and decreases with increasing temperature. For instance, the equilibrium concentration of ammonia at $T = 300\,°C$ and $p = 1\,000$ bar is as high as 93.39% whereas for $T = 600\,°C$ and $p = 50$ bar it amounts to the meager value of 2.28% only. In order to obtain a high amount of ammonia, it appears to be desirable to perform the synthesis at the highest possible pressure and the lowest possible temperature.

When Fritz Haber and Carl Bosch were developing the technology of ammonia synthesis at the beginning of the twentieth century, it turned out that high pressure was indeed beneficial for a high efficiency of the reaction. However, it also became apparent that the speed of reaction – a quantity that thermodynamics cannot predict – becomes prohibitively low at low temperatures. When the Badische Anilin and Soda Fabrik (BASF) in Oppau (Germany) started the first large-scale production of ammonia in the world in September 1913 using the Haber-Bosch method, one had chosen a pressure of approximately 100 bar and a temperature of 600 °C. The working temperature was a compromise between the need for a high concentration of ammonia (which required a low temperature) and the desire for a fast reaction (which required a high temperature). We should finally mention that the ideal gas model is insufficient to carry out realistic computations and that more complex equations of state are used in current industrial practice.

D – Suggestions for Further Study

Although Eq. 5.100 and Table 5.4 describe the dependence of the equilibrium concentration of ammonia on temperature and pressure quantitatively, none of these two representations is particularly appealing to physical intuition. To develop a better understanding of the equilibrium constant $K(T,p)$, it is advantageous to express its dependence on pressure in an explicit manner. To this end, we collect all pressure-dependent terms of the Gibbs functions g_A, g_B and g_C entering Eq. 5.101 and separate

Table 5.4. *Equilibrium states of the chemical reaction* $N_2 + 3H_2 \leftrightharpoons 2NH_3$. Concentration of ammonia n_C/n in percent for the case when the mole numbers of nitrogen and hydrogen are related to each other as 1:3, thereby forming a stochiometric mixture. The numbers in the table are not based on the simplified form (Eq. 5.92) of the Gibbs function but rather upon a more accurate albeit more complicated expression. The values have been taken from Appl (1999) with kind permission by Wiley-VCH Publisher, Weinheim, Germany

T (°C)	*p* (Mpa)										
	5	10	20	30	40	50	60	70	80	90	100
300	39.38	52.79	66.43	74.20	79.49	83.38	86.37	88.72	90.61	92.14	93.39
310	36.21	49.63	63.63	71.75	77.35	81.51	84.73	87.29	89.35	91.03	92.42
320	33.19	46.51	60.79	69.23	75.12	79.53	82.98	85.74	87.98	89.83	91.35
330	30.33	43.45	57.92	66.64	72.79	77.46	81.13	84.09	86.52	88.52	90.20
340	27.64	40.48	55.04	63.99	70.39	75.29	79.18	82.34	84.95	87.12	88.94
350	25.12	37.60	52.17	61.31	67.93	73.04	77.14	80.49	83.28	85.62	87.59
360	22.79	34.84	49.33	58.61	65.41	70.72	75.01	78.55	81.52	84.02	86.15
370	20.64	32.21	46.53	55.89	62.85	68.33	72.80	76.52	79.66	82.33	84.61
380	18.67	29.71	43.79	53.19	60.26	65.89	70.53	74.42	77.72	80.54	82.97
390	16.87	27.36	41.12	50.50	57.66	63.41	68.19	72.23	75.69	78.67	81.25
400	15.23	25.15	38.53	47.86	55.06	60.91	65.81	69.99	73.59	76.71	79.43
410	13.74	23.08	36.04	45.26	52.47	58.39	63.40	67.69	71.42	74.68	77.54
420	12.40	21.16	33.65	42.72	49.91	55.87	60.96	65.36	69.20	72.58	75.57
430	11.19	19.38	31.37	40.26	47.39	53.37	58.50	62.99	66.93	70.43	73.53
440	10.10	17.74	29.20	37.87	44.92	50.88	56.05	60.60	64.63	68.22	71.43
450	9.12	16.23	27.15	35.57	42.50	48.43	53.61	58.20	62.29	65.97	69.28
460	8.24	14.84	25.21	33.36	40.16	46.03	51.19	55.80	59.95	63.69	67.08
470	7.46	13.57	23.39	31.26	37.89	43.67	48.81	53.42	57.60	61.39	64.85
480	6.75	12.41	21.69	29.55	35.71	41.38	46.46	51.06	55.25	59.09	62.60
490	6.12	11.36	20.10	27.34	33.61	39.16	44.17	48.74	52.92	56.78	60.33
500	5.56	10.39	18.61	25.54	31.60	37.02	41.94	46.46	50.62	54.48	58.06
510	5.05	9.52	17.24	23.84	29.68	34.95	39.77	44.22	48.36	52.20	55.80
520	4.59	8.72	15.96	22.24	27.86	32.97	37.68	42.05	46.13	49.96	53.55
530	4.19	8.00	14.77	20.74	26.13	31.07	35.65	39.94	43.96	47.75	51.32
540	3.82	7.34	13.68	19.34	24.49	29.26	33.71	37.89	41.84	45.58	49.13
550	3.49	6.74	12.67	18.02	22.95	27.54	31.85	35.92	39.79	43.47	46.97
560	3.20	6.20	11.74	16.80	21.49	25.90	30.06	34.02	37.80	41.41	44.86
570	2.93	5.70	10.88	15.65	20.13	24.35	28.37	32.20	35.88	39.41	42.81
580	2.69	5.26	10.09	14.59	18.84	22.88	26.75	30.46	34.04	37.48	40.81
590	2.47	4.85	9.36	13.60	17.64	21.50	25.22	28.80	32.26	35.62	38.87
600	2.28	4.48	8.69	12.69	16.52	20.20	23.76	27.22	30.57	33.83	37.00

the exponential function into a pressure dependent and a temperature dependent part. After a straightforward computation which we leave as an exercise to the reader, we obtain

$$\frac{\left(\frac{n_C}{n}\right)^c}{\left(\frac{n_A}{n}\right)^a\left(\frac{n_B}{n}\right)^b} = p^{a+b-c}K_p(T) \tag{5.104}$$

This equation shows that for a chemical reaction with expansion ($a + b - c < 0$) the right-hand side decreases with increasing pressure and hence the concentration of reaction products decreases. Conversely, the concentration of reaction products increases if the chemical reaction is accompanied by a decreasing volume ($a + b - c > 0$) as it is the case for ammonia synthesis. (Remember that for given values of T and p the volume of one mole is the same for all gases. Therefore the stochiometric coefficients in the combination $a + b - c$ can be used to decide whether for a given reaction the volume increases or decreases.)

The function K_p depends on temperature only and can be considered to be an alternative form of the equilibrium constant. The reason for using the index "p" becomes clear if we express the numbers of moles through the partial pressures in the form $n_A/n = p_A/p$, $n_B/n = p_B/p$ and $n_C/n = p_C/p$. Then the law of mass action becomes

$$\frac{p_C^c}{p_A^a p_B^b} = K_p(T) \tag{5.105}$$

This formulation of the law of mass actions shows that the partial pressures of the gases are functions of the temperature only.

As a suggestion for further study, let us briefly address the temperature dependence of the equilibrium constant $K(T,p)$. Unfortunately it is not possible to express the temperature dependence in a form which is as simple as the pressure dependence in Eq. 5.103. However, by differentiation of $K(T,p)$ we can obtain the important relation

$$\left(\frac{\partial \ln K}{\partial T}\right)_p = \frac{\Delta h}{RT^2} \tag{5.106}$$

where Δh is the *heat of reaction*. For an exothermal process such as the Haber-Bosch reaction, $\Delta h < 0$. Thus the equilibrium constant decreases with increasing temperature in agreement with our observation in Table 5.4.

Finally we would like to emphasize that the law of mass action can be generalized to any number of reactants and reaction products. This can be verified by considering the example

$$a\mathrm{A} + b\mathrm{B} \rightleftharpoons c\mathrm{C} + d\mathrm{D} \tag{5.107}$$

This is a good occasion to become familiar with a notation which is often used in physical chemistry. Here the law of mass action is often expressed in terms of non-dimensional concentrations which are defined as

$$[\mathrm{A}] = \frac{n_\mathrm{A}/V}{1\ \mathrm{mol/l}} \tag{5.108}$$

For instance $[\mathrm{NH_3}] = 2$ stands for a concentration of ammonia of two moles per liter. If we use the ideal gas law to represent the quantities $n_\mathrm{A}/V, n_\mathrm{B}/V \ldots$ as $n_\mathrm{A}/V = p_\mathrm{A}/RT$, $n_\mathrm{B}/V = p_\mathrm{B}/RT \ldots$, Eq. 5.108 transforms into $p_\mathrm{A} = RT \cdot (1\ \mathrm{mol\,l^{-1}})[\mathrm{A}]$. Inserting this into Eq. 5.103 and adding a fourth substance D, a short calculation leads to

$$\frac{[\mathrm{C}]^c[\mathrm{D}]^d}{[\mathrm{A}]^a[\mathrm{B}]^b} = K_c(T) \tag{5.109}$$

The equilibrium constant K_c is related to K_p from Eq. 5.103 through

$$K_c(T) = (RT \cdot 1\ \mathrm{mol/l})^{a+b-c-d} K_p(T) \tag{5.110}$$

The three formulations 5.100, 5.103 and 5.105 of the law of mass action are physically identical. But the notation 5.105, which is popular with chemists, seems to be aesthetically most appealing. We therefore recommend the reader to memorize this form.

5.7 Production of Distilled Beverages

The art of distilling fermented grain, fruit or vegetables has evolved over hundreds of years without profound knowledge of the laws of thermodynamics. Whether we enjoy German "Nordhäuser Doppelkorn" from Thuringia, "Obstbrand" from the Black Forrest in Germany, Russian Vodka, French Cognac, Scotch Whisky or Chinese Luzhoulaojiaoerqujiu, it is thanks to the skills of experienced distillers who have passed their experiences on from generation to generation. However, thermodynamics has helped to produce these delicious beverages in large quantities and with uniform quality. In the present section, we wish to familiarize ourselves with the fundamentals of distillation for the understanding of which entropy plays a central role. The key to understanding distillation lies in the answer to the question of how it is possible to obtain a final product with a high content of alcohol from a raw material with a low content of alcohol. This question will be addressed in what follows.

A – Formulation of the Problem

We translate the question how to explain distillation into the language of thermodynamics by referring to Fig. 5.17. As a matter of fact, we must have recourse to a drastic simplification. We will omit all substances from our considerations except for water and alcohol and consider a thermodynamic system which consists of these two ingredients only. The exact chemical term for alcohol is ethanol. In the present work, however, we shall stick with the traditional term.

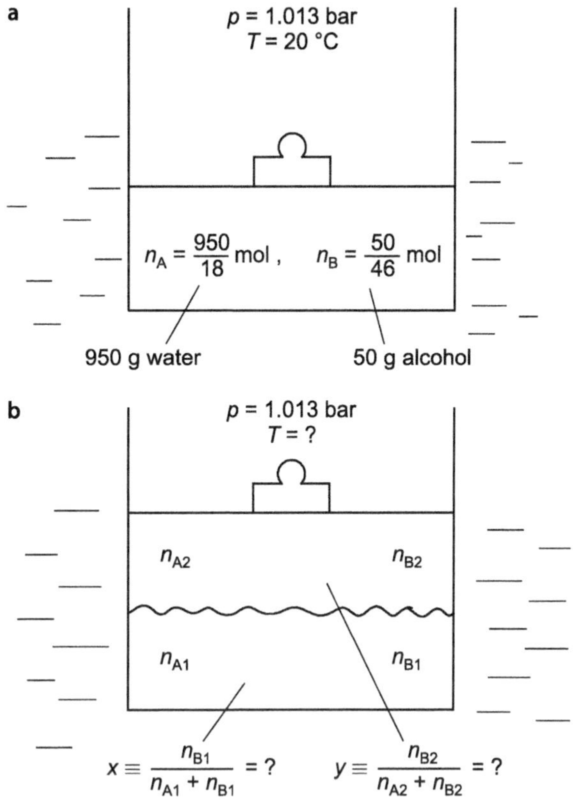

Fig. 5.17. *Concentration of alcohol during the production of distilled beverages.* (a) Initial state consisting of one kilogram of mash with an alcohol concentration of 5 mass-percent corresponding to a molar concentration of $x = n_B/(n_A + n_B) \approx 2\%$; (b) production state in which the mash boils and ejects a mixture of water vapor and alcohol vapor. Our goal is to determine the boiling temperature T as well as the mole fractions of the alcohol in the mash x and in the vapor y (In Sect. 5.7 E we will show that one of these three quantities has to be given in order to determine the other two quantities uniquely). The mole numbers n_A and n_B in (a) have been computed using the molar masses 18 g mol^{-1} for water and 46 g mol^{-1} for alcohol (C_2H_5OH). The pressure is assumed to be constant and to have the value $p_0 = 1.013$ bar

Distillation of liquors starts with a liquid with low alcohol content called mash. Let us assume that the mash is a homogeneous mixture of water and alcohol. To work with specific numbers, let us analyze 1 kg of mash with an alcohol content of 5 mass percent at a temperature of $T = 20$ °C and atmospheric pressure such as shown in Fig. 5.17. We assign the symbols A and B to the water and the alcohol, respectively. Then the mash contains $n_A = 52.7$ moles of water and $n_B = 1.09$ moles of alcohol.

The process of distillation consists in heating the mash until it boils. The emerging vapor consists of water vapor and alcohol vapor. Distillation is based on the fact that the alcohol content in the vapor is higher than in the mash. If the vapor flows through a cold pipe, it condenses into a liquid with a comparatively high content of alcohol. For instance in Whisky production, the product of the first step of distillation has a strength of 22% and is called "The Low" or "Low Wine." The process of boiling, which can be considered the production process, is visualiszd in Fig. 5.17b. Our thermodynamic system consists of two phases, namely a liquid and a gas, which we shall refer to as phases 1 and 2, respectively. Our goal is to predict the boiling temperature of our mixture as well as the concentrations x and y of alcohol in the liquid and in the vapor, respectively. x is defined as the ratio between the number of moles of liquid alcohol and the total number of moles in the liquid phase. y is the analogous quantity for the gaseous phase. In order to achieve our goal of computing x and y, we need to determine the entropy and the Gibbs function of a system consisting of two components (water and alcohol) and two phases (liquid and gas). We start by determining these quantities for the gaseous phase, then proceed with the liquid phase and finally summarize our findings for the whole system.

B – Entropy and Gibbs Function of a Dilute Mixture of Two Ideal Gases

In the previous section dealing with the production of ammonia, we have already derived the Expressions 5.86 and 5.90 for the entropy and Gibbs function of a mixture of ideal gases. We shall now simplify these formulas for the particular case of a *dilute mixture*. A dilute mixture is defined by the condition $n_B \ll n_A$ which expresses the fact that the mixture (for instance water vapor and alcohol vapor) contains only a very small amount of substance B (for instance alcohol). In the dilute case, the general expression for the mixing entropy

$$\Sigma(n_A, n_B) = -R\left(n_A \ln\frac{n_A}{n} + n_B \ln\frac{n_B}{n}\right) \tag{5.111}$$

can be simplified by making use of the Taylor expansions

$$\ln\left(\frac{n_A}{n_A + n_B}\right) = \ln\left(\frac{1}{1 + n_B/n_A}\right) \approx \ln\left(1 - \frac{n_B}{n_A}\right) \approx -\frac{n_B}{n_A} \tag{5.112}$$

$$\ln\left(\frac{n_B}{n_A + n_B}\right) = \ln\left(\frac{n_B/n_A}{1 + n_B/n_A}\right) \approx \ln\frac{n_B}{n_A} \tag{5.113}$$

The result is

$$\Sigma(n_A, n_B) \approx \Sigma_d(n_A, n_B) \equiv n_B \mathrm{R}\left(1 - \ln\frac{n_B}{n_A}\right) \tag{5.114}$$

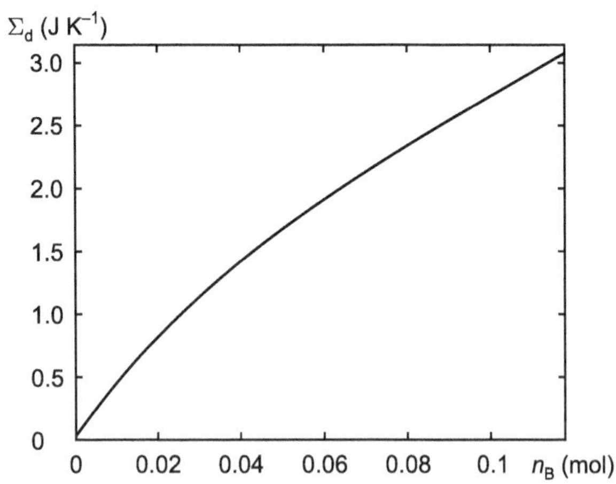

Fig. 5.18.
Behavior of the mixing entropy in the limit of very low concentration of one substance. In the case $n_B \ll n_A$ the mixing entropy can be approximated by the expression $\Sigma_d = n_B \mathrm{R}[1 - \ln(n_B/n_A)]$. The slope of the mixing entropy at $n_B = 0$ is infinite. (It is easy to over-salt a soup but difficult to desalinate an over-salted soup.) The curve is plotted for $n_A = 1$ mol

In deriving this expression, we have considered the quantity $\epsilon \equiv n_B / n_A$ as a small parameter and have used the Taylor series expansions $1/(1-\epsilon) \approx 1+\epsilon$ and $\ln(1-\epsilon) \approx -\epsilon$. The simplified mixing entropy given by Eq. 5.114 is marked with an index d, which reminds us that we are dealing with a dilute system. The dependence of this quantity on n_B is shown in Fig. 5.18.

Based on Eq. 5.86, we can write the entropy of a dilute mixture of two ideal gases in the symbolic form

$$S = n_A s_A(T,p) + n_B s_B(T,p) + \Sigma_d(n_A, n_B) \tag{5.115}$$

The explicit form of the entropy is given in Appendix E. We summarize our first result in the following way.

Entropy of a dilute mixture of two ideal gases. The entropy of a dilute ($n_B \ll n_A$) mixture of two ideal gases is equal to the sum of the entropies of the individual gases plus a quantity Σ_d which is independent of temperature and pressure, which depends only on the numbers of moles, and which has the form $\Sigma_d(n_A, n_B) = n_B R[1 - \ln(n_B / n_A)]$.

Based upon Eq. 5.90 we can write the Gibbs function $G = H - TS$ of our dilute mixture in the symbolic form

$$G = n_A g_A(T,p) + n_B g_B(T,p) - T\Sigma_d(n_A, n_B) \tag{5.116}$$

The explicit form of this expression is given in Appendix E. We can summarize this result as follows.

Gibbs function of a dilute mixture of two ideal gases. The Gibbs function of a dilute mixture of two ideal gases is equal to the sum of the Gibbs functions of the individual gases minus the term $T\Sigma_d$ where Σ_d is the mixing entropy for the particular case of a dilute system.

C – Entropy and Gibbs Function of a Dilute Ideal Solution

A liquid that consists of at least two miscible substances is called a *solution*. If the number of moles of substance A is much higher than of substance B, i.e., if $n_B \ll n_A$, the solution is referred to as a *dilute solution*. In a dilute solution, the substance A is called *solvent* and the substance B *solute*. A solution is called *ideal* if – simply speaking – its volume and internal energy are the sums of the volumes and internal energies of their constituents. If one adds 10 cm^3 of milk to a large pot of coffee, its volume increases by 10 cm^3. Hence, coffee with a small amount of milk is a good approximation to an ideal solution. By contrast, adding 10 cm^3 of alcohol to a large tank containing water leads to an increase of volume of only 9 cm^3. Hence, a mixture of water and alcohol is not a good representation of a model of an ideal solution. Since the development of a new theory should always start with the simplest possible case, we shall nevertheless assume in our present analysis that the mash is an ideal solution of water and alcohol.

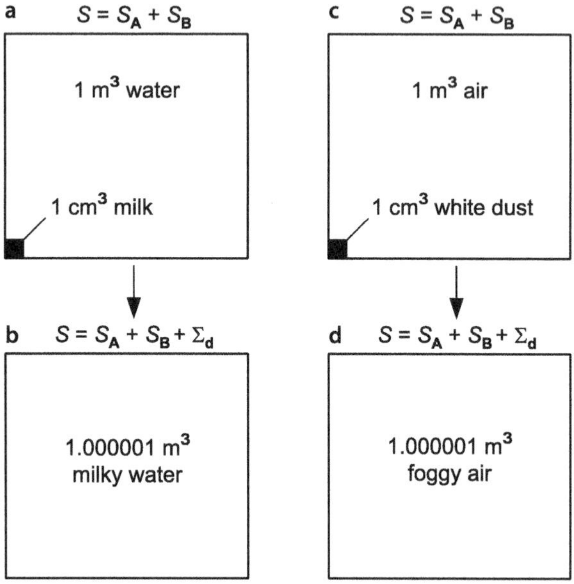

Fig. 5.19. *Mixing entropy for dilute solutions and gas mixtures.* An external observer cannot tell whether two liquids (a) form a dilute solution (b) or two gases (c) form a dilute gas mixture (d). Hence it is justified to approximate the mixing entropy of a dilute solution by the mixing entropy $\Sigma_d = n_B R[1 - \ln(n_B/n_A)]$ of a dilute mixture of gases

To compute the entropy of a dilute ideal solution, we start with the observation that the entropy of an ideal gas with constant (temperature-independent) heat capacity is

$$S = n\left[s_0 + c_p \ln(T/T_0) - R\ln(p/p_0)\right] \tag{5.117}$$

If the pressure p in the gas is constant and the normalization $p_0 = p$ is used, the entropy takes the form

$$S = n\left[s_0 + c\ln(T/T_0)\right] \tag{5.118}$$

where we have written c instead of c_p for simplicity. This expression is identical to the entropy of an incompressible substance which we have derived in Sect. 5.1. Hence the temperature dependence of the entropies of an ideal solution and a mixture of ideal gases is indistinguishable! As Fig. 5.19 illustrates, the mixing processes in an ideal solution and a gas are visually indistinguishable as well. It is thus tempting to describe the mixing entropy of a solution by using the same expression that we have used for ideal gases.

Following this hypothesis, we write the entropy of a dilute ideal solution in an analogous form as in Eq. 5.115, namely

$$S = n_A s_A(T) + n_B s_B(T) + \Sigma_d(n_A, n_B) \tag{5.119}$$

where s_A and s_B are the molar entropies of incompressible fluids. The full expression is given in Appendix E.

Some readers may not find the analogy leading to this equation convincing. To alleviate the concerns of those readers, we mention a thought experiment performed by Enrico Fermi and given in his thermodynamics textbook (Fermi 1956) which yields the same result and proceeds as follows. Since the mixing entropy depends only on the numbers of moles of the different substances but is independent of temperature and pressure, we can determine it by letting the temperature increase to such a high value that the whole solution evaporates. For sufficiently high temperatures, one can then regard the system as a mixture or two ideal gases whose mixing entropy we have computed both for the general case (Sect. 5.6 B) and for the particular case of a dilute mixture (Eq. 5.114). After having determined the mixing entropy in such a way, the gas mixture can be cooled down until it has condensed again without affecting the mixing entropy. Thus, one has effectively determined the mixing entropy for a dilute solution. We summarize our result as follows.

Entropy of a dilute ideal solution. The entropy of a dilute ($n_B \ll n_A$) ideal solution is equal to the sum of the entropies of its constituents plus a quantity Σ_d which is called mixing entropy and takes the form $\Sigma_d(n_A, n_B) = n_B R[1 - \ln(n_B/n_A)]$.

By virtue of the additivity of volume and internal energy, the enthalpy of the solution is

$$H = n_A\left[h_{0A} + c_A(T - T_0)\right] + n_B\left[h_{0B} + c_B(T - T_0)\right] \tag{5.120}$$

From this we readily deduce the Gibbs function $G = H - TS$ of the solution in the symbolic form

$$G = n_A g_A(T) + n_B g_B(T) - T\Sigma_d(n_A, n_B) \tag{5.121}$$

The explicit form of this expression is also given in Appendix E. We can summarize our result as follows.

Gibbs function of a dilute ideal solution. The Gibbs function of a dilute ideal solution is equal to the sum of the Gibbs functions of its constituents minus a term $T\Sigma_d$, where Σ_d is the mixing entropy for the particular case of a dilute solution.

D – Entropy and Gibbs Function of a Boiling Dilute Ideal Solution

We now turn to the final assembly of our theoretical tool. We collect Eqs. 5.115, 5.116, 5.119, and 5.121 for the entropy and for the Gibbs function of the gas and liquid, respectively. The result for the entropy is

$$\begin{aligned} S = {} & n_{A1}s_{A1}(T) + n_{B1}s_{B1}(T) + \Sigma_{d1}(n_{A1} + n_{B1}) \\ & + n_{A2}s_{A2}(T,p) + n_{B2}s_{B2}(T,p) + \Sigma_{d2}(n_{A2} + n_{B2}) \end{aligned} \tag{5.122}$$

In symbolic notation the Gibbs function becomes

$$\begin{aligned} G = {} & n_{A1}g_{A1}(T) + n_{B1}g_{B1}(T) + T\Sigma_{d1}(n_{A1} + n_{B1}) \\ & + n_{A2}g_{A2}(T,p) + n_{B2}g_{B2}(T,p) - T\Sigma_{d2}(n_{A2} + n_{B2}) \end{aligned} \tag{5.123}$$

The full explicit expressions for S and G are listed in Appendix E.

E – Result and Discussion

For given values of T and p and for fixed total numbers of moles n_A (water) and n_B (alcohol), our system contains two unknown parameters which describe the distribution of water and alcohol among the two phases. In analogy to the quantity ξ describing the extent of reaction during the production of ammonia (Sect. 5.6), we denote these parameters by ξ and η. We define them as

$$n_{A1} = (1-\xi)n_A, \quad n_{A2} = \xi n_A \tag{5.124}$$

$$n_{B1} = (1-\eta)n_B, \quad n_{B2} = \eta n_B \tag{5.125}$$

The Gibbs function $G(T,p,\xi,\eta)$ can now be regarded as dependent on these two parameters in addition to temperature and pressure. For instance, $\xi=0$ and $\eta=1$ would correspond to a state in which all water is liquid and all alcohol is gaseous. Strictly speaking, ξ and η do not represent thermodynamic coordinates, since they cannot be controlled from outside in contrast to the coordinates U and V. Instead, they attain certain values once the system has reached thermodynamic equilibrium, similar to the internal energies U_1 and U_2 of a thermal join discussed in Sect. 4.2 (see also Figs. 2.5 and 4.1). In the same way as it was useful in Sect. 4.2 to consider non-equilibrium states with a given total internal energy $U=U_1+U_2$ but different partitions U_1 and U_2, it is now instructive to compare the Gibbs functions for different values of ξ and η while keeping the total numbers of moles of water and alcohol constant. By incorporating these non-equilibrium states into our consideration, the dimension of our parameter space will temporarily increase from two ($X=(T,p)$) to four ($X=(T,p,\xi,\eta)$).

To find the equilibrium state of our system, we have to determine the minimum of the Gibbs function with respect to ξ and η while keeping temperature and pressure fixed. These conditions are

$$\frac{\partial G}{\partial \xi} = \left(\frac{\partial G}{\partial n_{A1}}\right)\frac{dn_{A1}}{d\xi} + \left(\frac{\partial G}{\partial n_{A2}}\right)\frac{dn_{A2}}{d\xi} = 0 \tag{5.126}$$

$$\frac{\partial G}{\partial \eta} = \left(\frac{\partial G}{\partial n_{B1}}\right)\frac{dn_{B1}}{d\eta} + \left(\frac{\partial G}{\partial n_{B2}}\right)\frac{dn_{B2}}{d\eta} = 0 \tag{5.127}$$

When computing the partial derivatives, we implicitly assume that during the differentiation all other quantities are kept constant. The derivatives of the numbers of moles with respect to ξ and η can be computed by using the definitions in Eq. 5.124. The results are $dn_{A1}/d\xi = -n_A$, $dn_{A2}/d\xi = n_A$, and similar expressions are readily derived for n_B. The equilibrium conditions then take the form

$$\frac{\partial G}{\partial n_{A1}} = \frac{\partial G}{\partial n_{A2}} \tag{5.128}$$

$$\frac{\partial G}{\partial n_{B1}} = \frac{\partial G}{\partial n_{B2}} \tag{5.129}$$

The derivative of the Gibbs function with respect to the number of moles is called the *chemical potential* and is often denoted by μ. The conditions 5.128 can be symbolically expressed as $\mu_{A1} = \mu_{A2}$ and $\mu_{B1} = \mu_{B2}$. They imply that the chemical potentials of a substance in the liquid and the gaseous phase must be equal. In addition to temperature and pressure, the chemical potential is the third quantity which describes an equilibration process, namely the equilibration of concentration.

Working out the derivatives (Eq. 5.128) for the Gibbs function given by Eq. 5.123 using the auxiliary relations $\partial\Sigma_d/\partial n_A = Rn_B/n_A$ and $\partial\Sigma_d/\partial n_B = -R\ln(n_B/n_A)$, we are led to the two equations

$$RT(x - y) = g_{A2}(T,p) - g_{A1}(T) \tag{5.130}$$

$$-RT(\ln x - \ln y) = g_{B2}(T,p) - g_{B1}(T) \tag{5.131}$$

Adopting standard notation we have replaced ξ and η with the *mole fractions* $x = n_{B1}/(n_{A1} + n_{B1})$ and $y = n_{B2}/(n_{A2} + n_{B2})$. For dilute systems $(n_{B1} \ll n_{A1})$ and $(n_{B2} \ll n_{A2})$ they simplify to $x = (1 - \eta)n_B/(1 - \xi)n_A$ and $y = \eta n_B/\xi n_A$. Equations 5.130 and 5.131 form a nonlinear system for the determination of the alcohol content of the mash $x(T,p)$ and for the alcohol content of the vapor $y(T,p)$. For our purpose, it is more convenient to regard the alcohol content of the mash x and the pressure p as given and to take Eqs. 5.130 and 5.131 as a system for determining the boiling temperature $T(p,x)$ as well as the alcohol content of the vapor $y(p,x)$.

It is impossible to solve this nonlinear system of equations analytically for the general case $x \in [0,1]$ and $y \in [0,1]$. However, since our expression for the Gibbs function is only valid for dilute systems $(x \ll 1, y \ll 1)$, it is sufficient to find an approximate solution valid for small mole fractions. We already know that for atmospheric pressure $p_0 = 1.013$ bar the boiling point of pure water is $T(p_0, 0) \approx 100$ °C. In other words, mash without alcohol boils like pure water. We therefore expand the unknown functions into a Taylor series for small mole fractions, use the abbreviation $T_0 = 100$ °C and write

$$T(p_0,x) = T_0 - \alpha x \tag{5.132}$$

$$y(p_0,x) = 0 + \beta x \tag{5.133}$$

Our task now reduces to determining the expansion coefficients α and β. After a lengthy calculation which is detailed in Appendix F, we obtain explicit expressions for α and β. We insert them into Eqs. 5.132 and 5.133 to obtain the desired result

$$T(p_0,x) = T_0 - \frac{RT_0^2}{\Delta h_A(T_0,p_0)}\left\{\exp\left[\frac{g_{B1}(T_0) - g_{B2}(T_0,p_0)}{RT_0}\right] - 1\right\}x \tag{5.134}$$

$$y(p_0,x) = \exp\left[\frac{g_{B1}(T_0) - g_{B2}(T_0,p_0)}{RT_0}\right]x \tag{5.135}$$

Here we have used the abbreviation $\Delta h_A(T_0, p_0) = T_0[s_{A2}(T_0, p_0) - s_{A1}(T_0)]$ for the enthalpy of evaporation of water. We now make use of the given value $x = 0.02$ (2%) for the initial alcohol concentration in the mash as well as of the thermodynamic data of pure water and pure alcohol. After a lengthy calculation which we will explain in detail in Appendix F, we obtain the final result

$$T = 99.6\,°\mathrm{C} \tag{5.136}$$

$$y = 0.0344 = 3.44\% \text{ (corresponding to 8.34 mass-\%)} \tag{5.137}$$

for the boiling temperature and the alcohol concentration of the vapor. This completes our analysis and explains the fundamentals of distillation which consist in the fact that the concentration of alcohol in the vapor (8.34%) is higher than in the liquid phase (5%).

Having proudly completed our analysis, let us now have a look at the quantitative accuracy of our theory. The experimental data for the water-alcohol system with a mole fraction of $x = 0.02$ (Landolt-Börnstein 1975, p. 38) are

$$T = 95.9\,°\mathrm{C} \tag{5.138}$$

$$y = 0.29 = 29\% \text{ (corresponding to 51 mass-\%)} \tag{5.139}$$

It turns out that the alcohol content in the vapor is much higher than predicted by our theory. Such a high alcohol concentration is certainly beneficial for producing good Whisky. Unfortunately, it also indicates that the predictive power of our theory is limited. The discrepancy between the prediction of our theory and the measurement bears witness to the fact that our description of a water-alcohol solution using the model of an ideal solution does not represent a very good approximation. The discrepancy also shows how important it is in chemical engineering to use accurate expressions for the entropy and Gibbs function of mixtures.

F – Suggestions for Further Study

After such tedious work, it is worth taking a look at the real phase diagram of the water-alcohol system which is shown in Fig. 5.20. The intersection of the dashed vertical line with the *boiling curve* $T(x)$ indicates the boiling temperature of the mash for a given mole fraction of alcohol x. The intersection of the dashed horizontal line with the curve $T(y)$ provides us with the alcohol content y in the gaseous phase. The effectiveness of the distillation of alcohol relies on the fact that the horizontal distance between the boiling curve and the curve describing the composition of the vapor is large. This permits one to obtain a vapor with comparatively high alcohol content after a single distillation step.

If we were to increase the alcohol content further, we would have to condense the vapor and distill the liquid condensate once again. If this process is repeated several times, as the dotted line in Fig. 5.20 illustrates, the technique is called multistep distillation. For further information about this process, we refer the reader to textbooks on chemical engineering. Diagrams as the one in Fig. 5.20 do not only play an important

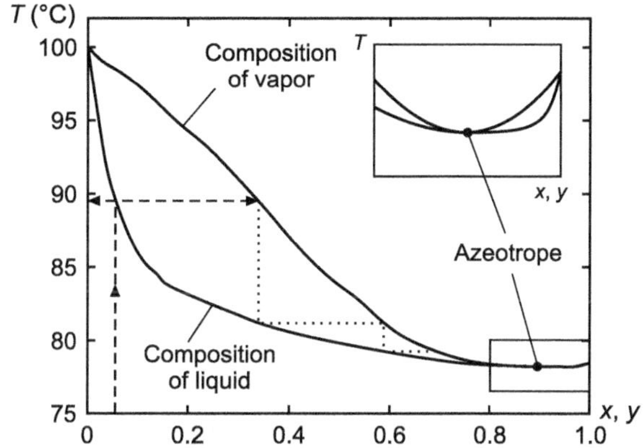

Fig. 5.20. *Fundamentals of Whisky production.* Phase diagram of the water-alcohol (ethanol) system. The diagram shows the boiling temperature $T(x)$ of a water-alcohol mixture as a function of the mole fraction x of alcohol in the liquid phase (labeled "composition of liquid") as well as the condensation temperature $T(y)$ of a mixture of water vapor and alcohol vapor as a function of the mole fraction y of alcohol in the gaseous phase (labeled "composition of vapor"). The theory of dilute ideal mixtures as used here describes approximately the slopes of the boiling and condensation curves in the vicinity of the point $x = y = 0$. The water-alcohol system is an azeotrope at $x = y \approx 0.96$ and $T \approx 78\,°\text{C}$ for which the liquid and the gaseous phases have the same composition and for which the boiling temperature is lower than that of both constituents. The numerical values have been taken from Gmehling and Onken (1977). The magnification is not to scale

role in describing phase transitions between liquids and gases but also for the coexistence of solids and liquids. For instance, the Fe-C phase diagram describes a variety of states of liquid and solid iron and is a very important tool in metallurgy.

The water-alcohol system exhibits one peculiarity, namely the existence of an azeotrope. An azeotrope is a mixture for which the liquid and the gaseous phases have the same composition. It follows from the existence of this point, which is shown in Fig. 5.20, that it is impossible to increase the alcohol content of an alcohol-water mixture beyond $x \approx 0.9$ by multistep distillation.

Let us finally look at our result (Eq. 5.134) from a more general perspective. Using Eq. A.26 from Appendix F, we can write Eq. 5.134 in the simplified form

$$T(p_0,x) = T_0 - \frac{RT_0^2}{\Delta h_A}\left[\exp\left(1 - \frac{p_0}{p_s}\right) - 1\right]x \tag{5.140}$$

$$y(p_0,x) = \exp\left(1 - \frac{p_0}{p_s}\right)x \tag{5.141}$$

It follows from this equation that if the saturation vapor pressure p_s of the solute at the boiling temperature T_0 of the pure solvent is larger than the vapor pressure of the solvent, then $y > x$ (the concentration of the solute in the gas is higher than in the liquid) and $T < T_0$ (the boiling temperature of the solution is lower than that of the pure solvent).

Equally interesting is the opposite case in which the solute has a lower vapor pressure than the solvent. An extreme example for such a situation is the solution of a solid, for instance a salt, in water. The vapor pressure of the solid is negligible which can be expressed as $p_s \rightarrow 0$. Hence $\exp(1 - p_0/p_s) \rightarrow 0$ and

$$T(p_0,x) = T_0 + \frac{RT_0^2 x}{\Delta h_A} \tag{5.142}$$

$$y(p_0,x) = 0 \tag{5.143}$$

The concentration of the solute in the gas phase is zero, and the boiling temperature increases. An analogous phenomenon is the melting point depression. This effect is responsible for the observation that sea water has a solidification point below 0 °C.

Chapter 6 Summary

Summary

In the present book, we have gone a long way from the mathematical foundations of entropy to its practical application. Now we shall look back onto our main achievements and summarize our conclusions with the following ten sentences.

1. The equilibrium states of a thermodynamic system can be sorted using the order relation $\prec$ which is called adiabatic accessibility;
2. A state Y is called adiabatically accessible from a state X, in symbols $X \prec Y$ (pronounced "X precedes Y") if it is possible to change the state from X to Y by means of an interaction with some device and a weight, in such a way that the device returns to its initial state at the end of the process whereas the weight may have changed its position in a gravitational field (Lieb and Yngvason 1999);
3. Our physical experience and mathematical assumptions about the adiabatic accessibility of equilibrium states of thermodynamic systems can be condensed into 15 axioms of the order relation $\prec$;
4. Without any further assumptions, in particular without using terms like temperature and heat, it is possible to derive the following entropy principle from these axioms from which the conclusions 6–10 follow in turn;
5. **Entropy principle.** For each equilibrium state X of a thermodynamic system there is a real valued function $S(X)$, called entropy. The entropy is **monotonic** [$X \prec\prec Y$ implies $S(X) < S(Y)$, and $X \overset{A}{\sim} Y$ implies $S(X) = S(Y)$], **additive** [$S((X, Y)) = S(X) + S(Y)$], extensive [$S(tX) = tS(X)$] and **concave** [$S(tX + (1-t)Y) \geq tS(X) + (1-t)S(Y)$];
6. An adiabatic process with the initial state X and the final state Y is called irreversible if $S(X) < S(Y)$ and reversible if $S(X) = S(Y)$;
7. For each simple system with the entropy $S(U, V)$ (where U is the internal energy and V one or several work coordinates), there is an intensive property $T = (\partial U/\partial S)_V$ called temperature. Two systems are in thermal equilibrium if their temperatures are equal;
8. The quantity $Q = \int T(S)\mathrm{d}S$ defined for a quasistatic process is called heat;
9. Clausius' formulation of the second law of thermodynamics: No process is possible; the sole result of which is that heat is transferred from a body to a hotter one;
10. Kelvin and Planck's formulation of the second law of thermodynamics: No process is possible; the sole result of which is that a body is cooled and work is done.

Finally, let us take a look at the questions formulated at the beginning of Chap. 1. Our analysis in Sect. 5.1 has answered the questions regarding the use of the thermal energy of the Gulf Stream, and Sect. 5.2 provides an answer to the problem of producing ice. With our acquired knowledge, it is also not difficult to answer the remaining questions.

The flow of a glacier is due to the fact that the high pressure on its lower boundary lets the ice melt. We have explained this phenomenon in connection with ice skating using the Clausius-Clapeyron equation in Sect. 5.3. However, the pressure dependence of the melting temperature of ice is not the only reason for the flow of a glacier. This flow is in fact the result of a complicated interplay of the melting point depression and the nonlinear rheological properties of ice.

The death of a diver is due to the pressure-dependence of the solubility of air in blood. During diving at great depths, the thermodynamic equilibrium state of the two-component two-phase system consisting of air and blood (cf. Sect. 5.7) is shifted towards a high nitrogen concentration in the blood. If the diver rises, the pressure decreases and the solubility of nitrogen in blood decreases as well. Hence the gas is expelled from the blood in a similar way to carbon dioxide being released from a bottle of soda. This inflicts destruction upon the organs and the diver dies. The thermodynamics of this tragic event can in principle be described using the same mathematical methods as in the process of producing distilled beverages, discussed in Sect. 5.7. In the simplest case, we would describe the blood-air system as a two-component system and determine the Gibbs functions of both the liquid phase consisting of water and air and the gaseous phase consisting of water vapor and air. By minimizing the Gibbs function for given values of temperature and air concentration in the liquid phase, we could compute the concentration of air $x(p)$ as a function of pressure. For more information, we refer the reader to Enrico Fermi's textbook on thermodynamics (Fermi 1956).

The problem of foam formation in cola is similar to the previously discussed problem. The solubility of carbon dioxide is a monotonically decreasing function of temperature. One could analyze the formation of foam in cola using the same method as just described. Considering cola as a mixture of water and carbon dioxide involving a liquid and a gaseous phase, we would have to write down the Gibbs function of the total system. By minimizing the Gibbs function for a given pressure, we would obtain the equilibrium concentration of carbon dioxide $x(T)$ as a function of the temperature. This function would decrease with increasing temperature, as predicted. We should emphasize once more that entropy is crucial for the solution of this problem because the existence of the Gibbs function relies on the existence of entropy. Without entropy this problem would be intractable.

Let us finally discuss the question why salt extracts water from a cucumber. This is related to the phenomenon of *osmosis*. If one separates salt water from fresh water using a membrane which is impermeable for salt but permeable for the water, the pressure in the saltwater will increase. This pressure difference is called the osmotic pressure. This pressure can be calculated by minimizing the Helmholtz function of the system. Details can be found in Fermi's textbook on thermodynamics (Fermi 1956). But why does osmosis explain the behavior of a salted cucumber? This question is left to the reader as a parting gift.

References and Further Reading

References and Further Reading

References

Appl M (1999) Ammonia – principles and industrial practice. Wiley-VCH, Weinheim

Avery WH, Wu C (1994) Renewable energy from the ocean: A guide to OTEC. Oxford University Press, Oxford

Carathéodory C (1909) Untersuchung über die Grundlagen der Thermodynamik. Mathematische Annalen 67:355–386

Carnot S (1824, reprinted 1990) Réflexions sur la puissance motrice du feu et sur les machines propres a développer cette puissance. éditions Jacques Gabay, Sceaux (English translation: Fox R (1986) Reflexions on the motive power of fire. Manchester University Press)

Callen H (1985) Thermodynamics and an introduction to thermostatistics. John Wiley & Sons, New York

Engemann S, Reichert H, Dosch H, Bilgram J, Honkimäki V, Snigirev A (2004) Interfacial melting of ice in contact with SiO_2. Phys Rev Lett 92:205701

Feynman RP, Leighton RB, Sands M (1963) The Feynman lectures on physics, vol. I. Addison Wesley, Reading

Fermi E (1956) Thermodynamics. Dover Publications, New York

Gmehling J, Onken U (1977) Chemistry data series, vol. I, part 1: Vapor-liquid equilibrium data collection. DECHEMA, Frankfurt

Giles R (1964) Mathematical foundations of thermodynamics. Pergamon, Oxford

Landolt HH, Börnstein R (1974) Landolt-Börnstein: Zahlenwerte und Funktionen aus Naturwissenschaften und Technik, Band 3 – Thermodynamisches Gleichgewicht siedender Gemische. Springer-Verlag, Berlin

Lieb EH, Yngvason J (1999) The physics and mathematics of the second law of thermodynamics. Phys Rep 310:1–96

Lieb EH, Yngvason J (2000) A fresh look at entropy and the second law of thermodynamics. Phys Today 4:32–37

Moran MJ, Shapiro HN (1995) Fundamentals of engineering thermodynamics. John Wiley & Sons, New York

Planck M (1964) Vorlesungen über Thermodynamik. Walter de Gruyter & Co., Berlin (in German)

Rosenberg R (2005) Why is ice slippery? Phys Today 12:50–55

Suggestions for Further Reading

This book is based on the article by Lieb and Yngvason (1999) which is freely available on the internet. This article contains the formulation of the axioms given in Appendix B, the derivation of the entropy principle sketched in Chap. 3 as well as proofs of the conclusions from the entropy principle that we have discussed in Chap. 4. This article is remarkable for its accuracy, elegance and logic and can be considered the bible of thermodynamics. Its technical level is far above the present book. Nevertheless, we strongly recommend all readers to take a look at this article. Even those readers who are not interested in mathematical proofs will certainly find it useful and entertaining to read the introduction and summary. The same authors have published a short popular version of their article (Lieb and Yngvason 2000) in the journal *Physics Today*. This article can be downloaded from the internet for free, albeit without figures. The level of difficulty of this popular article is below the level of this book. It is therefore suitable as an introduction to the present work.

Some readers may wish to compare the Lieb-Yngvason formulation of thermodynamics with traditional derivations of entropy and the second law via Carnot cycles. We recommend that these readers consult two classical textbooks, namely chapters 44 to 46 of "*The Feynman Lectures on Physics*" (Feynman et al. 1963), which is famous for its clarity and "*Thermodynamics*" (Fermi 1956) which does not only cover the fundamentals of thermodynamics but also a number of applications such as dilute solutions, reacting gas mixtures and the law of mass action. A presentation of thermodynamics which stands between the traditional treatment and the Lieb-Yngvason formulation is given in the textbook by Callen (1985). This book enjoys great popularity among physicists. Finally, for an extensive treatment of practical applications of thermodynamics including vapor power systems, refrigeration systems, gas turbines, internal combustion engines and moist air we refer the reader to the textbook on engineering thermodynamics by Moran and Shapiro (1995).

Historically interested readers are referred to "*Réflexions sur la puissance motrice du feu et sur les machines propres a développer cette puissance*" by Sadi Carnot (1824) (English translation 1986, see references), in which the universal efficiency of Carnot cycles is demonstrated. This work is a milestone in the historical development of thermodynamics. However, viewed from today's perspective, its logic is somewhat difficult to understand. Also of historical interest is the article by Carathéodory (1909) which was the first attempt to put thermodynamics on a solid mathematical basis. Carathéodory's work can be viewed as a foundation of Lieb and Yngvason's work. Finally, we should mention the book by Giles (1964), which represents the closest precursor of the Lieb-Yngvason theory.

Appendices

Appendices

Appendix A
Hans in Luck

by the brothers Grimm,
translated by Robert Goodwin-Jones,
reproduced with kind permission

Hans had served his master for seven years, so he said to him, "Master, my time is up. Now I should be glad to go back home to my mother, so please give me my wages."

The master answered, "You have served me faithfully and honestly. As the service was, so shall the reward be." And he gave Hans a piece of gold as big as his head.

Hans pulled his handkerchief out of his pocket, wrapped up the lump in it, put it on his shoulder, and set out on the way home.

As he went on, always putting one foot before the other, he saw a horseman trotting quickly and merrily by on a lively horse. "Ah," said Hans quite loudly, "what a fine thing it is to ride. There you sit as if on a chair, you stumble over no stones, you save your shoes, and cover the ground, you don't know how."

The rider, who had heard him, stopped and called out,

"Well, then, Hans, why do you go on foot?"

"I must," answered he, "for I have this lump to carry home. It is true that it is gold, but it is so heavy that I cannot hold up my head, and it hurts my shoulder."

"I will tell you what," said the rider, "we will make an exchange. I will give you my horse, and you can give me your lump of gold."

"With all my heart," said Hans, "but I can tell you, you will have the weary task of crawling along with it."

The rider got down, took the gold, and helped Hans up, then put the bridle securely in his hands and said, "If you want to go at a really good pace, you must click your tongue and call out, jup, jup."

Hans was heartily delighted as he sat upon the horse and rode away so boldly and freely. After a little while he thought that it ought to go faster, and he began to click with his tongue and call out "jup, Jup!"

The horse put himself into a sharp trot, and before Hans knew where he was, he was thrown off and lying in a ditch which separated the field from the highway. The horse would have gone off too if it had not been stopped by a countryman, who was coming along the road and driving a cow before him. Hans pulled himself together and stood up on his legs again.

He was vexed, and said to the countryman, "It is a poor joke, this riding, especially when one gets hold of a mare like this that kicks and throws one off, so that one has a chance of breaking one's neck. Never again will I mount it. Now I like your cow, for one can walk quietly behind her, and have, over and above, one's milk, butter and cheese every day without fail. What would I not give to have such a cow."

"Well," said the countryman, "if it would give you so much pleasure, I do not mind giving the cow for the horse."

Hans agreed with the greatest delight, so the countryman jumped upon the horse and rode quickly away.

Hans drove his cow quietly before him and thought over his lucky bargain. "If only I had a morsel of bread – and that can hardly fail me – I could eat butter and cheese with it as often as I liked, and whenever I am thirsty, I can milk my cow and drink the milk. My goodness, what more could I ever want."

When he came to an inn he made a halt, and in his great concern he ate up what he had with him – his dinner and supper – and all he had, and with his last few farthings he had half a glass of beer. Then he drove his cow onwards along the road to his mother's village.

As it drew nearer to midday, the heat was more oppressive, and Hans found himself upon a moor which it took about an hour to cross. He felt it very hot and his tongue clave to the roof of his mouth with thirst.

"I can find a cure for this," thought Hans. "I will milk the cow now and refresh myself with the milk."

He tied her to a withered tree, and as he had no pail he put his leather cap underneath, but try as he would, not a drop of milk came. And as he set himself to work in a clumsy way, the impatient beast at last gave him such a blow on his head with its hind foot that he fell on the ground, and for a long time he could not remember where he was.

By good fortune a butcher soon came down the road with a wheelbarrow carrying a young pig.

"What sort of a trick is this," cried he, and helped the good Hans up. Hans told him what had happened. The butcher gave him his flask and said, "take a drink and refresh yourself. The cow will certainly give no milk; it is an old beast, at best it is only fit for the plough or for the butcher."

"Well, well," said Hans, as he stroked his hair down on his head, "who would have thought it. Certainly it is a fine thing when one can kill a beast like that at home, what meat one would have. But I do not care much for beef; it is not juicy enough for me. A young pig like that now is the thing to have; it tastes quite different, and then there are the sausages."

"Listen, Hans," said the butcher, "out of love for you I will make an exchange and let you have the pig for the cow."

"Heaven repay you for your kindness," said Hans as he gave up the cow, whilst the pig was unbound from the barrow, and the cord by which it was tied was put in his hand.

Hans went on and thought to himself how everything was going just as he wished; if he did meet with any vexation it was immediately set right. Presently there joined him a lad who was carrying a fine white goose under his arm. They said good morning to each other, and Hans began to tell of his good luck, and how he had always made such good bargains. The boy told him that he was taking the goose to a christening feast.

"Just lift her," added he, and took hold of her by the wings, "how heavy she is – she has been fattened up for the last eight weeks. Whosoever has a bit of her when she is roasted will have to wipe the fat from both sides of his mouth."

"Yes," said Hans, as he weighed her in one hand, "she is a good weight, but my pig is no bad one." Meanwhile the lad looked suspiciously from one side to the other, and he shook his head.

"Look here," he said at length, "it may not be all right with your pig. In the village through which I passed, the mayor himself had just had one stolen out of its sty. I fear – I fear that you may have got hold of it there. They have sent out some people to search for it, and it would be a bad business if they caught you with the pig; at the very least, you would be shut up in the dark hole."

The good Hans was terrified. "Goodness," he said, "help me out of this fix! You know more about this place than I do. Take my pig and leave me your goose."

"I shall risk something at that game," answered the lad, "but I will not be the cause of your getting into trouble."

So he took the cord in his hand, and drove away the pig quickly along a bypath. The good Hans, free from care, went homewards with the goose under his arm.

"When I think over it properly," said he to himself, "I have even gained by the exchange. First there is the good roast meat, then the quantity of fat which will drip from it, and which will give me dripping for my bread for a quarter of a year, and lastly the beautiful white feathers. I will have my pillow stuffed with them, and then indeed I shall go to sleep without rocking. How glad my mother will be."

As he was going through the last village, there stood a scissors-grinder with his barrow. As his wheel whirred he sang,

I sharpen scissors and quickly grind,
My coat blows out in the wind behind.

Hans stood still and looked at him; at last he spoke to him and said, "All's well with you, as you are so merry with your grinding."

"Yes," answered the scissors-grinder, "the trade has a golden foundation. A real grinder is a man who as often as he puts his hand into his pocket finds gold in it. But where did you buy that fine goose?"

"I did not buy it, but exchanged my pig for it."

"And the pig?"

"That I got for a cow."

"And the cow?"

"I took that instead of a horse."

"And the horse?"

"For that I gave a lump of gold as big as my head."

"And the gold?"

"Well, that was my wages for seven years of service."

"You have known how to look after yourself each time," said the grinder. "If you can only get on so far as to hear the money jingle in your pocket whenever you stand up, you will have made your fortune."

"How shall I manage that?" said Hans.

"You must be a grinder, as I am. Nothing particular is wanted for it but a grindstone, the rest finds itself. I have one here, and it is certainly a little worn, but you need not give me anything for it but your goose, will you do it?"

"How can you ask," answered Hans. "I shall be the luckiest fellow on Earth. If I have money whenever I put my hand in my pocket, why should I ever worry again?" And he handed him the goose and received the grindstone in exchange.

"Now," said the grinder, as he took up an ordinary heavy stone that lay by him, "here is a strong stone for you into the bargain; you can hammer well upon it, and straighten your old nails. Take it with you and keep it carefully."

Hans loaded himself with the stones, and went on with a contented heart, his eyes shining with joy. "I must have been born with a caul," he cried, "for everything I want happens to me just as if I were Sunday's child."

Meanwhile, as he had been on his legs since daybreak, he began to feel tired. Hunger also tormented him, for in his joy at the bargain by which he got the cow he had eaten up all his store of food at once. At last he could only go on with great trouble, and was forced to stop every minute; the stones, too, weighed him down dreadfully. Then he could not help thinking how nice it would be if he had not to carry them just then.

He crept like a snail to a well in a field, and there he thought that he would rest and refresh himself with a cool draught of water, but in order that he might not injure the stones in sitting down, he laid them carefully by his side on the edge of the well. Then he sat down on it, and was to stoop and drink, when he made a slip, pushed against the stones, and both of them fell into the water.

When Hans saw them with his own eyes sinking to the bottom, he jumped for joy, and then he knelt down with tears in his eyes to thank God for having shown him this favor also, and delivering him in so good a way, and without his having any need to reproach himself, from those heavy stones which had been the only things that troubled him.

"There is no man under the sun as fortunate as I," he cried out.

With a light heart and free from every burden, he now ran on until he was with his mother at home.

Appendix B
Axioms for the Derivation of the Entropy Principle

Here we give the full set of axioms which are necessary to derive the entropy principle. The axioms are taken from Lieb and Yngvason's original work (Lieb and Yngvason 1999) and will be neither explained nor discussed. We give them here in order to summarize the fundamentals of thermodynamics and to encourage the readers to study the original paper.

General Axioms

A-1 Reflexivity. $X \overset{A}{\sim} X$.

A-2 Transitivity. $X \prec Y$ and $Y \prec Z$ implies $X \prec Z$.

A-3 Consistency. $X \prec X'$ and $Y \prec Y'$ implies $(X, Y) \prec (X', Y')$.

A-4 Scaling invariance. If $X \prec Y$, then $tX \prec tY$ for all $t > 0$.

A-5 Splitting and recombination. For $0 < t < 1$, $X \overset{A}{\sim} (tX, (1-t)X)$.

A-6 Stability. If $(X, \epsilon Z_0) \prec (Y, \epsilon Z_1)$ holds for a sequence of ϵ tending to zero and some states Z_0 and Z_1, then $X \prec Y$.

A-7 Convex combination. Assume X and Y are states in the same state space, Γ, that has a convex structure. If $t \in [0, 1]$, then $(tX, (1-t)Y) \prec tX + (1-t)Y$.

Axioms for Simple Systems

Let Γ, a convex subset of $\boldsymbol{R}^{n+1}$ for some $n > 0$, be the state space of a simple system.

S-1 Irreversibility. For each $X \in \Gamma$ there is a point $Y \in \Gamma$ such that $X \prec\prec Y$.

S-2 Lipschitz tangent planes. For each $X \in \Gamma$, the forward sector $A_X = \{Y \in \Gamma : X \prec Y\}$ has a unique support plane at X (i.e., A_X has a tangent plane at X). The slope of the tangent plane is assumed to be a locally Lipschitz continuous function of X.

S-3 Connectedness of the boundary. The boundary ∂A_X of a forward sector is connected.

Axioms for Thermal Equilibrium

T-1 Thermal contact. For any two simple systems with state spaces Γ_1 and Γ_2, there is another simple system, the *thermal join* of Γ_1 and Γ_2, with state space

$$\Delta_{12} = \{(U, V_1, V_2) : U = U_1 + U_2 \quad \text{with } (U_1, V_1) \in \Gamma_1 \quad \text{and } (U_2, V_2) \in \Gamma_2\} \qquad \text{(A.1)}$$

Moreover,

$$\Gamma_1 \times \Gamma_2 \ni ((U_1,V_1),(U_2,V_2)) \prec (U_1 + U_2,V_1,V_2) \in \Delta_{12} \tag{A.2}$$

T-2 Thermal splitting. For any point $(U_1, V_1, V_2) \in \Delta_{12}$ there is at least one pair of states, $(U_1, V_1) \in \Gamma_1$, $(U_2, V_2) \in \Gamma_2$ with $U = U_1 + U_2$ such that

$$(U,V_1,V_2) \overset{A}{=} ((U_1,V_1),(U_2,V_2)) \tag{A.3}$$

In particular, if (U, V) is a state of a simple system and $\lambda \in [0, 1]$, then

$$(U,(1-\lambda)V,\lambda V) \overset{A}{=} ((1-\lambda)U,(1-\lambda)V,(\lambda U,\lambda V)) \in \Gamma^{(1-\lambda)} \times \Gamma^{(\lambda)} \tag{A.4}$$

If $(U, V_1, V_2) \overset{A}{\sim} ((U_1, V_1), (U_2, V_2))$ we write $(U_1, V_1) \overset{T}{\sim} (U_2, V_2)$.

T-3 Zeroth law. If $X \overset{T}{\sim} Y$ and if $Y \overset{T}{\sim} Z$ then $X \overset{T}{\sim} Z$.

T-4 Transversality. If Γ is the state space of a simple system and if $X \in \Gamma$, then there exist states $X_0 \overset{T}{\sim} X_1$ with $X_0 \prec\prec X \prec\prec X_1$.

T-5 Universal temperature range. If Γ_1 and Γ_2 are state spaces of simple systems, then for every $X \in \Gamma_1$ and every V in the projection of Γ_2 onto the space of its work coordinates, there is a $Y \in \Gamma_2$ with work coordinates V such that $X \overset{T}{\sim} Y$.

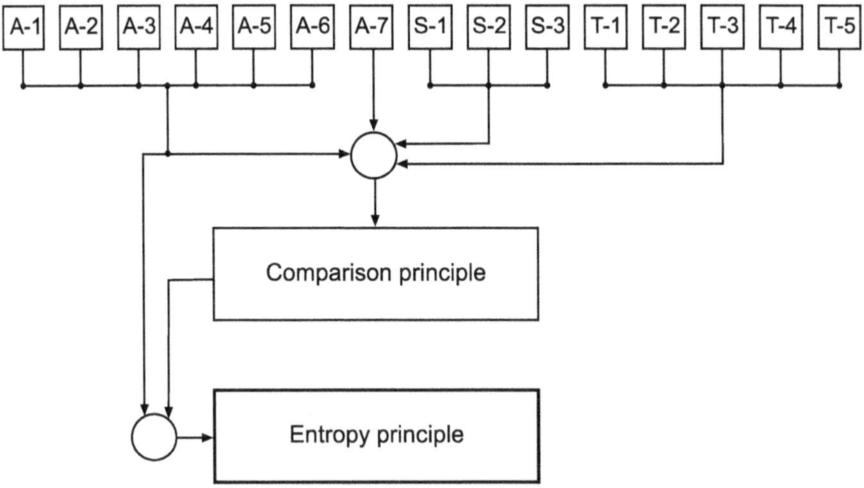

Fig. A.1. *The logical structure of the entropy principle.* Overview of the relation between the axioms of thermodynamics, the comparison principle and the entropy principle. The entropy principle can be derived from the axioms A-1 through A-6 and the comparison principle. All axioms are necessary to derive the comparison principle. The mixing axiom is not included in the figure. For details see Lieb and Yngvason (1999)

Axiom for Mixtures and Reactions

Two state spaces, Γ and Γ', are said to be connected, written $\Gamma \prec \Gamma'$, if there are state spaces $\Gamma_0, \Gamma_1, \Gamma_2, \ldots, \Gamma_N$ and states $X_i \in \Gamma_i$ and $Y_i \in \Gamma_i$, for $i = 1, \ldots, N$ and states $\tilde{X} \in \Gamma$ and $\tilde{Y} \in \Gamma'$ such that $(\tilde{X}, X_0) \prec Y_1$, $X_i \prec Y_{i+1}$ for $i = 1, \ldots, N-1$ and $X_N \prec (\tilde{Y}, Y_0)$.

M **Absence of sinks.** If Γ is connected to Γ', then Γ' is connected to Γ, i.e., $\Gamma \prec \Gamma' \Longrightarrow \Gamma' \prec \Gamma$.

Comparison Hypothesis

The logical connection between the axioms and the entropy principle (Sect. 3.4) is represented in Fig. A.1. The comparison hypothesis, which is also given in Fig. A.1, has the following form:

> **Comparison hypothesis (CH).** The comparison hypothesis is said to hold for a state space if any two states X and Y in the space are comparable, i.e., $X \prec Y$ or $Y \prec X$.

Appendix C
Irreversible and Reversible Heat Transfer

In many technical processes, heat is transferred from a reservoir to a working substance such as ammonia in Sect. 5.4. (Remember, a reservoir is a simple thermodynamic system whose work coordinate does not change.) In general, heat transfer is an irreversible process as illustrated in Fig. A.2 and as shown next.

Let us assume that a reservoir with temperature T_H transfers the energy Q_H to a working substance. Let us further assume that the pressure of the substance is held constant and that the substance transforms from liquid to gaseous state in response to the added heat. The consideration of the special case of a two-phase system makes our computation particularly simple and easy to follow. During the evaporation, the temperature T does not change. Before we can analyze whether this process is irreversible or reversible, we have to check whether the process is adiabatic. Indeed, as we have shown in Sect. 4.1, the definition of reversibility can only be given for adiabatic processes. In the present case, we have a compound system which consists of the reservoir and the working substance. Since no change occurs outside the system, we are indeed dealing with an adiabatic process.

As we have demonstrated in Sect. 5.1 C, the entropy of a reservoir decreases by Q_H/T_H when the energy Q_H is extracted from it. In deriving this formula, we have assumed that the reservoir is so large that its temperature does not change during the heat transfer process. At the same time, the entropy of the working substance increases by $S_g - S_f$ when it evaporates. Hence the entropy of the compound system changes by $\Delta S = -Q_H/T_H + (S_g - S_f)$. According to Sect. 5.3 (Eq. 5.38) the entropy is a linear function of the enthalpy in the two-phase region. Hence we can write the entropy difference as

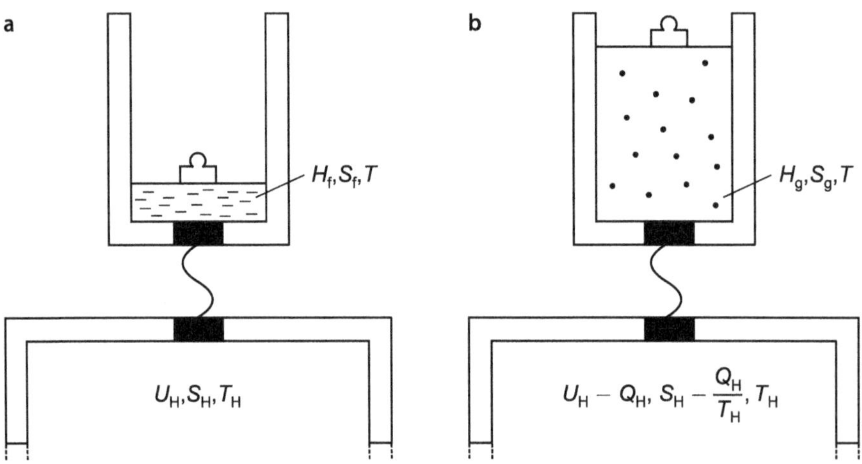

Fig. A.2. *Irreversible and reversible heat transfer.* Illustration of entropy production associated with heat transfer. When the energy Q_H is extracted from an infinitely large reservoir with the temperature T_H and used to evaporate a liquid at the temperature T, this process is irreversible in general. In the limit $T_H \to T$, however, the process is reversible and proceeds infinitely slow. The two systems have been chosen such that they do not change their temperatures during the process of heat transfer

$S_g - S_f = (H_g - H_f) / T$. Due to energy conservation, the enthalpy change during evaporation is $H_g - H_f = Q_H$. It follows that the change of entropy of the compound system is

$$\Delta S = \left(\frac{1}{T} - \frac{1}{T_H}\right) Q_H \qquad \text{(A.5)}$$

Since heat is always transferred from high to low temperature, the expression inside the brackets is always positive. It follows that the change of entropy is positive, in agreement with the law of entropy increase. Consequently, heat transfer is an irreversible process in general.

But we can also see in this equation that the produced entropy becomes smaller if the temperature difference is decreased. In the limit of a vanishing temperature difference ($T \rightarrow T_H$), the entropy production tends to zero. This limiting case is referred to as isothermal heat transfer. Isothermal heat transfer is a reversible process! As noted in Feynman et al. (1963), isothermal heat transfer is the thermal analog to frictionless movement in classical mechanics. However, since the heat flux also tends to zero when $T \rightarrow T_H$, isothermal heat transfer proceeds infinitely slowly and represents an idealized concept.

Appendix D
Properties of the Mixing Entropy

Here we list several properties of the mixing entropy which help the reader to develop an intuitive understanding of this quantity. The mixing entropy of an ideal gas consisting of n_A moles of gas A and n_B moles of gas B is

$$\Sigma(n_A,n_B) = -R\left(n_A \ln\frac{n_A}{n} + n_B \ln\frac{n_B}{n}\right) \tag{A.6}$$

with $n = n_A + n_B$. It has the following properties:

- it is positive, $\Sigma \geq 0$;
- it is a monotonically increasing function of the mole numbers n_A and n_B;
- it is extensive, i.e., $\Sigma(\lambda n_A, \lambda n_B) = \lambda\Sigma(n_A, n_B)$;
- $\Sigma = 0$ if either $n_A = 0$ or $n_B = 0$;
- if $n_A = n_B = n/2$ then the mixing entropy depends upon the mole number as $\Sigma = nR\ln 2$.

Where does the name *mixing* entropy come from? Figure A.3 helps to understand the answer. The figure shows two gases which are enclosed in a vessel at a given temperature T and pressure p. In the initial state, they are separated by a thin wall. In this state, the entropy of the system is obviously given by the sum

$$S_1 = n_A s_A(T,p) + n_B s_B(T,p) \tag{A.7}$$

Once the walls are removed, the gases will mix. In the final state, a homogeneous mixture is formed. Earlier, we derived Formula 5.87

$$S_2 = n_A s_A(T,p) + n_B s_B(T,p) + \Sigma(n_A,n_B) \tag{A.8}$$

for this mixture. The difference $S_2 - S_1$ is equal to the mixing entropy. Since the mixing entropy is positive and mixing is an adiabatic process, the mixing of two gases is an irreversible process.

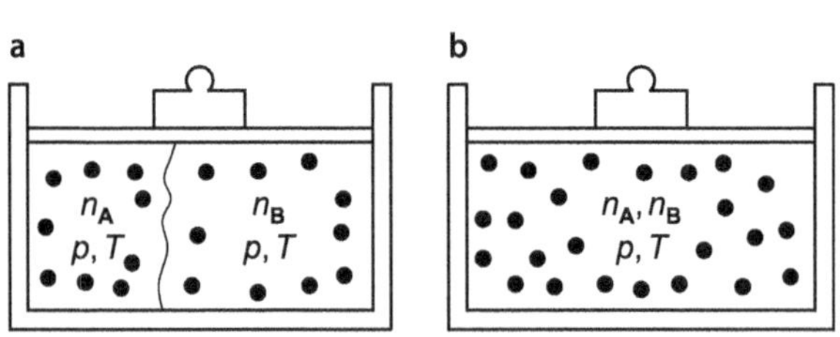

Fig. A.3. *Interpretation of the mixing entropy.* Illustration of the mixing process of two ideal gases. (a) Initial state; (b) final state. The entropies of the initial and the final state differ by a quantity that is called mixing entropy. In the initial state, the two gases are separated by a membrane

Appendix E
Entropy and Gibbs Function of a Dilute Mixture of Ideal Gases and of a Dilute Ideal Solution

The entropy of a dilute ($n_B \ll n_A$) mixture of two ideal gases is symbolically described by Eq. 5.115. The full expression is

$$\begin{aligned} S(T,p,n_A,n_B) = {} & n_A\left[s_{0A} + c_{pA}\ln(T/T_0) - R\ln(p/p_0)\right] \\ & + n_B\left[s_{0B} + c_{pB}\ln(T/T_0) - R\ln(p/p_0)\right] \\ & + Rn_B\left[1 - \ln(n_B/n_A)\right] \end{aligned} \tag{A.9}$$

The Gibbs function, whose symbolic expression is given in Eq. 5.116, has the explicit form

$$\begin{aligned} G(T,p,n_A,n_B) = {} & n_A\left\{h_{0A} + c_{pA}(T - T_0) - T\left[s_{0A} + c_{pA}\ln(T/T_0) - R\ln(p/p_0)\right]\right\} \\ & + n_B\left\{h_{0B} + c_{pB}(T - T_0) - T\left[s_{0B} + c_{pB}\ln(T/T_0) - R\ln(p/p_0)\right]\right\} \\ & - RTn_B\left[1 - \ln(n_B/n_A)\right] \end{aligned} \tag{A.10}$$

The short form of the entropy of a dilute ($n_B \ll n_A$) ideal solution is given by Eq. 5.119. The corresponding full expression is

$$\begin{aligned} S(T,n_A,n_B) = {} & n_A\left[s_{0A} + c_A\ln(T/T_0)\right] \\ & + n_B\left[s_{0B} + c_B\ln(T/T_0)\right] \\ & + Rn_B\left[1 - \ln(n_B/n_A)\right] \end{aligned} \tag{A.11}$$

The Gibbs function (Eq. 5.121) has the explicit form

$$\begin{aligned} G(T,n_A,n_B) = {} & n_A\left\{h_{0A} + c_{pA}(T - T_0) - T\left[s_{0A} + c_A\ln(T/T_0)\right]\right\} \\ & + n_B\left\{h_{0B} + c_{pB}(T - T_0) - T\left[s_{0B} + c_B\ln(T/T_0)\right]\right\} \\ & - RTn_B\left[1 - \ln(n_B/n_A)\right] \end{aligned} \tag{A.12}$$

The entropy of a system consisting of two components A and B, for instance water and alcohol, and two phases 1 and 2, for instance liquid and gas, has been given in short form in Eq. 5.122. The explicit expression is

$$\begin{aligned} S(T,p,n_{A1},n_{A2},n_{B1},n_{B2}) = {} & n_{A1}\left[s_{0A1} + c_A\ln(T/T_0)\right] \\ & + n_{B1}\left[s_{0B1} + c_B\ln(T/T_0)\right] \\ & + Rn_{B1}\left[1 - \ln(n_{B1}/n_{A1})\right] \\ & + n_{A2}\left[s_{0A2} + c_{pA}\ln(T/T_0) - R\ln(p/p_0)\right] \\ & + n_{B2}\left[s_{0B2} + c_{pB}\ln(T/T_0) - R\ln(p/p_0)\right] \\ & + Rn_{B2}\left[1 - \ln(n_{B2}/n_{A2})\right] \end{aligned} \tag{A.13}$$

Finally, the symbolic notation of the Gibbs function (Eq. 5.123) has the explicit form

$$
\begin{aligned}
&G(T,p,n_{A1},n_{A2},n_{B1},n_{B2}) \\
&\quad = n_{A1}\left\{h_{0A1} + c_A(T - T_0) - T\left[s_{0A1} + c_A \ln(T/T_0)\right]\right\} \\
&\quad + n_{B1}\left\{h_{0B1} + c_B(T - T_0) - T\left[s_{0B1} + c_B \ln(T/T_0)\right]\right\} \\
&\quad - RTn_{B1}\left[1 - \ln(n_{B1}/n_{A1})\right] \\
&\quad + n_{A2}\left\{h_{0A2} + c_{pA}(T - T_0) - T\left[s_{0A2} + c_{pA} \ln(T/T_0) - R\ln(p/p_0)\right]\right\} \\
&\quad + n_{B2}\left\{h_{0B2} + c_{pB}(T - T_0) - T\left[s_{0B2} + c_{pB} \ln(T/T_0) - R\ln(p/p_0)\right]\right\} \\
&\quad - RTn_{B2}\left[1 - \ln(n_{B2}/n_{A2})\right] \qquad (A.14)
\end{aligned}
$$

This equation is the basis for the analysis of the production of distilled beverages.

Appendix F
Auxiliary Expressions for the Analysis of the Production of Distilled Beverages

After inserting the formulations 5.132 and 5.133 into Eqs. 5.130 and 5.130, we expand the right-hand sides in Taylor series up to the first order. The result is

$$g_{A1}(T) \approx g_{A1}(T_0) - \left(\frac{\partial g_{A1}}{\partial T}\right)_{T=T_0} \alpha x = g_{A1}(T_0) + s_{A1}(T_0)\alpha x \tag{A.15}$$

$$g_{A2}(T, p_0) \approx g_{A2}(T_0, p_0) - \left(\frac{\partial g_{A2}}{\partial T}\right)_{T=T_0} \alpha x = g_{A2}(T_0, p_0) + s_{A2}(T_0, p_0)\alpha x \tag{A.16}$$

Analogous relations hold for g_{B1} and g_{B2}. In deriving these formulas, we have made use of $\partial g / \partial T = -s$, which follows from the equation $\mathrm{d}G = -S\mathrm{d}T + V\mathrm{d}p$ for the Gibbs function. Inserting the foregoing expansions into Eq. 5.130, we obtain

$$\mathrm{R}(T_0 - \alpha x)(\beta - 1)x = g_{A2}(T_0, p_0) + s_{A2}(T_0, p_0)\alpha x - g_{A1}(T_0) - s_{A1}(T_0)\alpha x \tag{A.17}$$

The left-hand side of this relation contains terms of both first and second order in x. If we neglect second-order terms, the left-hand side takes the form $\mathrm{R}T_0(\beta - 1)x$. The first and third terms on the right-hand side cancel since the Gibbs function remains constant during a phase transition. The difference $s_{A2}(T_0, p_0) - s_{A1}(T_0)$ is equal to the enthalpy of evaporation of water Δh_A divided by T_0. Hence Eq. A.17 takes the form

$$\beta - 1 = \frac{\alpha \Delta h_A}{\mathrm{R}T_0^2} \tag{A.18}$$

Next we take the relation for the Gibbs functions of the alcohol which is analogous to Eq. A.15. We insert it into Eq. 5.131 and obtain

$$-R(T_0 - \alpha x)\ln\beta = g_{B2}(T_0, p_0) + s_{B2}(T_0, p_0)\alpha x - g_{B1}(T_0) - s_{B1}(T_0)\alpha x \tag{A.19}$$

This equation contains terms of both first and second order in x. By comparing the coefficients in front of the zero-order terms, we obtain

$$-RT_0 \ln\beta = g_{B2}(T_0, p_0) - g_{B1}(T_0) \tag{A.20}$$

However, the differences between the Gibbs functions appearing here do not disappear. This is due to the fact that T_0 is not equal to the boiling temperature of pure alcohol at p_0! If we solve this equation for β and insert the result into Eq. A.18, we obtain the desired solution as

$$\alpha = \frac{\mathrm{R}T_0^2}{\Delta h_A}\left\{\exp\left[\frac{g_{B1}(T_0) - g_{B2}(T_0, p_0)}{\mathrm{R}T_0}\right] - 1\right\} \tag{A.21}$$

$$\beta = \exp\left[\frac{g_{B1}(T_0) - g_{B2}(T_0, p_0)}{RT_0}\right] \tag{A.22}$$

Inserting this into Eq. 5.132 yields the final result of Eq. 5.134.

If all parameters of the entropy and Gibbs functions are known, namely h_{0A1}, h_{0A2}, h_{0B1}, h_{0B2}, s_{0A1}, s_{0A2}, s_{0B1}, s_{0B2}, c_A, c_B, c_{pA} and c_{pB}, the quantities α and β can be computed. They in turn permit one to compute the boiling temperature and the alcohol concentration in the gaseous phase as functions of x. In practice, the mentioned parameters are not tabulated in this form. Instead, it is customary to tabulate data for pure water and alcohol such as enthalpies, entropies and specific volumes as functions of temperature or vapor pressure. For instance, a table for pure water gives an evaporation enthalpy of 2 257 kJ kg^{-1} at $T_0 = 100$ °C from which we derive a molar enthalpy of evaporation of

$$\Delta h_A = 4.104 \times 10^4 \text{ J mol}^{-1} \tag{A.23}$$

For pure alcohol at $T_0 = 100$ °C, the vapor pressure is

$$p_s = 2.184 \text{ bar} \tag{A.24}$$

These two figures are sufficient to compute the desired quantities.

We can write the Gibbs function of the gaseous alcohol in the following way:

$$\begin{aligned} g_{B2}(T_0, p_0) &= g_{B2}(T_0, p_s + (p_0 - p_s)) \\ &\approx g_{B2}(T_0, p_s) + \left(\frac{\partial g}{\partial p}\right)_{p=p_s} (p - p_s) \\ &= g_{B2}(T_0, p_s) + v_{B2}(T_0, p_s)(p_0 - p_s) \\ &\approx g_{B2}(T_0, p_s) + \frac{RT_0}{p_s}(p_0 - p_s) \end{aligned} \tag{A.25}$$

The further analysis proceeds in three steps. First, we use a Taylor series expansion. Second, we use the relation $\partial g / \partial p = v$. Third, we approximate the specific volume $v_{B2}(T_0, p_s)$ by the equation of state of an ideal gas. Then the difference, which appears in Eq. A.21, takes the form

$$g_{B1}(T_0) - g_{B2}(T_0, p_0) = g_{B1}(T_0) - g_{B2}(T_0, p_s) - RT_0\left(1 - \frac{p_0}{p_s}\right) \tag{A.26}$$

The first two terms on the right-hand side cancel because the Gibbs function does not change during a phase transition. Hence

$$\frac{g_{B1}(T_0) - g_{B2}(T_0, p_0)}{RT_0} = \left(1 - \frac{p_0}{p_s}\right) \tag{A.27}$$

Appendix G
Explanation of the Examples of Entropy Production in Everyday Life

Quenching a Boiled Egg

We consider the egg as an incompressible substance with mass $m = 70$ g and specific heat capacity $c = 4\,184$ J kg^{-1} K^{-1} equal to that of water. We further assume that the water in the saucepan is a reservoir whose mass is much higher than that of the egg and whose temperature $T_C = 10$ °C is constant. During the quenching process, the egg cools from its initial temperature $T_H = 100$ °C to $T_C = 10$ °C and thereby reduces its internal energy by $Q = mc(T_H - T_C)$. The internal energy of the cold water increases by the same amount. According to Eq. 5.6, the entropy of the egg decreases by

$$\sigma_H = mc\ln\frac{T_H}{T_C} = 80.84\ \text{J}\,\text{K}^{-1} \tag{A.28}$$

In Sect. 5.1 we have derived Eq. 5.8 for the change of entropy of a reservoir. Using this expression we obtain

$$\sigma_C = \frac{Q}{T_C} = mc\left(\frac{T_H}{T_C} - 1\right) = 93.11\ \text{J}\,\text{K}^{-1} \tag{A.29}$$

for the increase of entropy of the cold water. The change of entropy of the compound system consisting of the egg and the water is thus

$$\Delta S = \sigma_C - \sigma_H = 12.27\ \text{J}\,\text{K}^{-1} \tag{A.30}$$

Explosion of a Tire

We consider a compound system consisting of the air within the tire and of the surrounding air and assume that this system undergoes an adiabatic process. Suppose that the total kinetic energy W of the rubber pieces flying through the air is represented by increasing the potential energy of a weight by the amount W. We assume that the air in the tire is an ideal gas with mass $m = 20$ g. Given the effective molar mass of air of 28.9 g mol^{-1}, this is equivalent to considering $n = 1.730$ moles of air. During the blast, the air is decompressed from its initial pressure $p = 2.2$ bar to atmospheric pressure $p_0 \approx 1$ bar. The temperature of the air in the tire is equal to the ambient temperature T_0 both in the initial and the final state. (Observe that during the explosion, the system is not in thermodynamic equilibrium and no temperature can be defined in general. If the blast would proceed through a succession of equilibrium states, the temperature would in general be different from T_0.) According to Eq. 5.24, the entropy of the air in the tire increases by

$$\sigma_R = nR\ln\frac{p}{p_0} = 11.34\ \text{J}\,\text{K}^{-1} \tag{A.31}$$

By contrast, its internal energy remains unchanged. This is because the internal energy of an ideal gas depends on temperature only and the temperatures are the same for the initial and the final states. Hence energy conservation requires that the internal energy of the ambient air has decreased by the same amount as the work W imparted to the weight by our compound system. Since we have assumed that the ambient air is a reservoir (remember that a reservoir is a system with only one thermodynamic coordinate, the energy coordinate), and its temperature remains unchanged, its entropy must have decreased by

$$\sigma_A = \frac{W}{T_0} \tag{A.32}$$

The entropy of the compound system therefore changes by

$$\Delta S = n\mathrm{R}\ln\frac{p}{p_0} - \frac{W}{T_0} \tag{A.33}$$

According to the law of entropy increase, the change of entropy after an adiabatic process obeys $\Delta S \geq 0$. From this we conclude that the work performed in the course of the tire explosion must satisfy the inequality $W \leq n\mathrm{R}T_0 \ln(p/p_0)$. The maximum possible work is obtained if we replace the inequality with an equality. Since $\Delta S = 0$, this case corresponds to a reversible process. However, a reversible explosion of a tire is not realistic because it would have to proceed infinitely slowly. In the opposite case, $W = 0$, the tire performs no work. Then the entropy production according to Eq. A.33 is maximum and takes the value $n\mathrm{R}\ln(p/p_0) = 11.34\ \mathrm{J\,K^{-1}}$. We thus arrive at the conclusion that the entropy produced as a result of an exploding tire is confined to the interval

$$0 \leq \Delta S \leq 11.34\ \mathrm{J\,K^{-1}} \tag{A.34}$$

Pouring Water over Heated Rocks in the Sauna

We consider $m = 300$ g water with specific heat capacity $c = 4\,184\ \mathrm{J\,kg^{-1}\,K^{-1}}$ and initial temperature $T_C = 20$ °C which comes in contact with heated rocks having a constant temperature of $T_H = 200$ °C. The water instantly heats up and evaporates as soon as it has reached a temperature of $T = 100$ °C. We shall assume that the mass of the rocks is so large in comparison with m that their temperature remains virtually unchanged during the evaporation process.

The computation of the entropy increase of the water proceeds in two steps. First, we use Eq. 5.6 for the entropy of an incompressible substance to determine the entropy increase of the liquid water when it is heated from 20 °C to 100 °C. The result is

$$\sigma_{C1} = mc\ln\frac{T}{T_C} = 302.8\ \mathrm{J\,K^{-1}} \tag{A.35}$$

Second, we invoke Eq. 5.41 for the entropy of a liquid-gas mixture to determine the entropy increase due to evaporation. The application of this formula results in

$$\sigma_{C2} = \frac{m\Delta h}{T} = 1815 \text{ J K}^{-1} \tag{A.36}$$

where $\Delta h = 2\,257$ kJ kg^{-1} is the enthalpy of evaporation of water under atmospheric pressure. The entropy increase of the water is therefore $\sigma_C = \sigma_{C1} + \sigma_{C2} = 2\,118$ J K^{-1}.

The entropy of the rocks decreases by

$$\sigma_H = \frac{\Delta H}{T_H} = 1643 \text{ J K}^{-1} \tag{A.37}$$

where $\Delta H = mc(T_H - T_C) + m\Delta h = 777.5$ kJ is an abbreviation for the decrease of the enthalpy of the rocks (ΔH is also equal to the increase of the enthalpy of the water). The entropy production in the compound system consisting of the water and the rocks is finally

$$\Delta S = \sigma_H - \sigma_C = 475 \text{ J K}^{-1} \tag{A.38}$$

Distribution of Perfume

We consider $m_P = 1$ g of perfume after it has been vaporized. We would like to determine the entropy production when the perfume distributes uniformly within a room. We assume that the room contains $m_A = 50$ kg of air which corresponds to approximately 50 m^3. For simplicity we shall assume that the perfume consists of pure alcohol and has a molar mass of $M_P = 46$ g mol^{-1}. The effective molar mass of air is $M_A = 28.9$ g mol^{-1}. Given the masses and the molar masses of both the alcohol and air, we readily compute their mole numbers as $n_P = 0.02174$ mol and $n_A = 1\,730$ mol. The indices P and A stand for perfume and air, respectively. To compute the difference between the entropies in the initial and final states, we need to compute the entropy of the unmixed gases according to Eq. 5.84 and compare it with the entropy of the mixture described by Eq. 5.86. Since we assume that both the perfume and the air behave as ideal gases, their temperature and pressure remain constant during the mixing process. Hence, the terms containing temperature and pressure cancel and the only contribution to the entropy difference comes from the mixing entropy Σ. Taking into account that the total number of moles is $n = n_P + n_A$ we thus obtain the entropy production as

$$\Delta S = \Sigma(n_P, n_A) = -\mathrm{R}\left(n_P \ln\frac{n_P}{n} + n_A \ln\frac{n_A}{n}\right) = 2.220 \text{ J K}^{-1} \tag{A.39}$$

In deriving this result we have ignored the entropy produced during the vaporization process. Therefore, this figure represents a lower bound for the entropy production.

Dissolution of Salt in Water

We consider $m_S = 1$ g of salt which dissolves in $m_W = 1$ kg water and forms a homogeneous solution. We ignore the entropy production connected with the process of solution of the salt and focus on the process of mixing, similar to the previous example.

Hence the entropy production computed below represents a lower bound on the true entropy production. The molar mass of salt (NaCl) is 58.45 g mol^{-1}. We therefore have $n_S = 0.01712$ mol of salt. The molar mass of water is 18 g mol^{-1}, and its mole number is 55.55 mol. Since $n_S \ll n_W$, we have a dilute mixture. As in the previous example, we assume that the temperature and pressure remain constant during the mixing process. Hence the entropy production can be computed from Eq. 5.119 from which we can remove the terms depending on temperature and pressure. The only contribution comes from the mixing entropy Σ_d (where the index d refers to the dilute case) and we obtain

$$\Delta S = \Sigma_d(n_S, n_W) = -Rn_S\left(1 - \ln\frac{n_S}{n_W}\right) = 1.293\ \mathrm{J\,K^{-1}} \tag{A.40}$$

Index

M

N

O

P

Q

R

S

T

GPSR Compliance

The European Union's (EU) General Product Safety Regulation (GPSR) is a set of rules that requires consumer products to be safe and our obligations to ensure this.

If you have any concerns about our products, you can contact us on ProductSafety@springernature.com

In case Publisher is established outside the EU, the EU authorized representative is:

Springer Nature Customer Service Center GmbH
Europaplatz 3
69115 Heidelberg, Germany

Batch number: 10682215

Printed by Printforce, the Netherlands